Making Medicine Scientific

Making Medicine Scientific

John Burdon Sanderson and the Culture of Victorian Science

Terrie M. Romano

The Johns Hopkins University Press

Baltimore and London

This book has been brought to publication with the generous
assistance of a grant from the Social Sciences and Humanities
Research Council of Canada.

The Johns Hopkins University Press
2715 North Charles Street
Baltimore, Maryland 21218-4363
www.press.jhu.edu

Library of Congress Cataloging-in-Publication Data

Romano, Terrie M.
Making medicine scientific : John Burdon Sanderson and the
culture of Victorian science / Terrie M. Romano.
p.   cm.
"In memory of Bernardo Romano (1938–1989) and
Saveria Carlone Romano (1939–1983)."
Includes bibliographical references and index.
ISBN 0-8018-6897-1
1. Burdon-Sanderson, J. (John), Sir, 1828–1905.   2. Physicians—
Great Britain—Biography.   3. Medicine—Great Britain—
History—19th century.   4. Medical sciences—Great Britain—
History—19th century.   I. Title.
R489.B875 R66 2002
610'.92—dc21   2001004375

A catalog record for this book is available
from the British Library.

In memory of Bernardo Romano (1938–1989)
and Saveria Carlone Romano (1939–1983)

Contents

Acknowledgments

While working on this book, I have accumulated many debts, more than I can fully describe here. Sandra McRae encouraged me when I most needed it and interested me in the subject of scientific medicine. Robert Frank mentioned in a footnote that no one had studied John Burdon Sanderson. I am most deeply indebted to those who read and commented on earlier drafts of the entire manuscript: Sandra den Otter, Jacalyn Duffin, Alison Li, Charles Rosenberg, Molly Sutphen, an anonymous reviewer for Johns Hopkins University Press, and Jacqueline Wehmueller. Their suggestions and encouragement were of enormous help.

I would like to thank Larry Holmes and John Harley Warner for their insightful criticisms throughout my time at Yale University and for their comments on several drafts of this project in its earlier incarnations. My revisions also benefited from the criticisms of the late Gerald L. Geison.

Many scholars have been generous with their time. I thank Christopher Lawrence, Stephen Jacyna, and William Summers for many stimulating discussions over the years. I want to thank Dorothy Porter, Steve Sturdy, and Frank Turner for their advice and criticism, and I am especially grateful to Linda Colley for her incisive comments. Mark Curthoys was an invaluable guide to the University of Oxford, and Michael Barfoot kindly shared his insights into the University of Edinburgh with me. I thank Mark Curthoys, Jacalyn Duffin, John Parascondola, Margaret Pelling, Carolyn Shapiro, Nancy Slack, James Strick, and Maria Trumpler for drawing information and sources to my attention which I might otherwise have overlooked. Patty Pauls and Chris Beck of Queen's Medical Photography quickly produced slides on many occasions; Robert Sullivan redesigned figure 2.3 at a moment's notice. I also thank Pat Johnson, Joanna Gorman, and Cherrilyn Yalin for their valuable assistance at different stages of the project. I would like to thank everyone at the Johns Hopkins University Press who helped me, especially Jacqueline Wehmueller, Lee Sioles, and Elizabeth Gratch.

Early on the Wellcome Institute for the History of Medicine provided me with a supportive environment in London. I am grateful to W. F. Bynum for his practical help in Britain and for many fruitful discussions. Later I benefited from my time as a visiting scholar at the Wellcome Unit at the University of Cambridge, for which I would particularly like to thank Andrew Cunningham and Harmke Kamminga. I have enjoyed speaking about John Burdon Sanderson and Victorian medical science to various audiences in Canada, the United States, and Great Britain. All of their reactions have helped shape this book.

Many archivists and librarians facilitated my research. First, I would like to thank Ferenc Gyorgyey of the Medical Historical Library, Yale University, for his help. In Britain I want to thank all those who assisted me at the University College London Archives; the Wellcome Library, London, and the Contemporary Medical Archives Centre; the National Library of Scotland; the University of Edinburgh Archives; the Greater London Record Office; the Public Record Office; the Bodleian Library; and the British Library. I am also grateful to Anne Barrett at Imperial College London, Peter Donavan at the Marylebone Reference Library, Simon Bailey at the Oxford University Archives, and Tony Simcock at the Museum of the History of Science, Oxford. In North America I thank Lee Perry, Woodward Library, University of British Columbia, and Elizabeth Tunis and Stephen Greenberg, National Library of Medicine. The Library, University College London; The Bodleian Library, Oxford; and The Wellcome Trust, London, all kindly granted me permission to reproduce images and to quote from documents in their possession.

I thank the editors of the *Bulletin of the History of Medicine* and Johns Hopkins University Press and the editors of the *Journal of the History of Medicine and Allied Sciences* and Oxford University Press for permission to reprint portions of my articles "Gentlemanly versus Scientific Ideals: John Burdon Sanderson, Medical Education, and the Failure of the Oxford School of Physiology" and "The Cattle Plague of 1865 and the Reception of 'the Germ Theory' in Mid-Victorian Britain," both of which originally appeared in 1997.

I acknowledge with gratitude the essential funding by Yale University, the Hannah Institute for the History of Medicine, Queen's University, and the Social Sciences and Humanities Research Foundation of Canada.

The hospitality of Sarah Divall and Elli Glevey, Jackie Nawka and Vartan Armeniakan, and Nick and Marie Nawka in London made my research possible and more enjoyable. Finally, I would like to acknowledge Kay Magaard, whom I met as I began this project. He and our two sons, Lorenz and Carlo, did not directly help me with this book, but they share in my joy at its completion.

Making Medicine Scientific

Introduction

In 1829 Tertius Lydgate, a gentleman and surgeon, moved to the English provincial town of Middlemarch. In this, the period leading up to the great Reform Act of 1832, change was in the air, and Lydgate saw himself as a reformer of the medical profession. Imbued with the spirit of scientific culture, "he was fired with the possibility that he might work out the proof of an anatomical discovery." Lydgate was not alone. Although he was the fictional creation of George Eliot for her 1870 novel *Middlemarch,* Lydgate represented the many young, ambitious medical practitioners of the first half of the nineteenth century who had studied in Paris (and later in German medical centers) and returned to Britain and America with some knowledge of medical research. They began their careers full of enthusiasm for science and with the desire to add to medical knowledge. Lydgate was also typical because he failed. His life followed a familiar pattern, and in the end his early research aspirations were engulfed by intellectual isolation and by his need to make a living through his practice.

This book is centered on the career of Sir John Scott Burdon Sanderson (1829–1905) who, like the fictional Lydgate, was a gentleman who aimed to make medicine scientific. Unlike Lydgate, Burdon Sanderson successfully and atypically made a career for himself as a medical researcher in the second half of the nineteenth century. In many ways a prototypical Victorian, Burdon Sanderson grew up in an Evangelical family, began his career as a medical practitioner and medical officer of health in London, went on to do pathological and physiological research, first in London and then at the University of Oxford, and ended his career as Regius Professor of Medicine at Oxford.

This book explores the many meanings of scientific medicine, and how an ambitious medic could become a professional researcher in Victorian Britain. In short, scientific medicine was supported and opposed because it was a complex of ideologies and practices which meant different things to different people. This study illustrates the importance of the Victorian milieu—religious, political, and philosophical—to the campaign for a scientific medicine. Burdon Sanderson's career also highlights the existence of "biological" scientific researchers in Victorian Britain apart from the scientific naturalists of T. H. Huxley's circle. Interested in reforming medical practice, they belonged more to the clinical and pathological community. Underpinning the narrative is the importance of pathology in garnering allies for laboratory medicine in the British context and the disorderly emergence of pathology and physiology as separate fields out of pathological anatomy. The campaign for scientific medicine must be understood in a Victorian setting. The tangible, material context included the dirt and disease of the era as well as the material culture of experimentation—from frogs to photographs. Turning to the intangible, philosophical context, the proponents of scientific medicine drew on a belief in science and strains of Evangelicalism, positivism, and idealism.

As an admirer of Xavier Bichat (1771–1802), Lydgate's idea of proof was that of the pathological anatomy school of Paris, which described illness as localized lesions of the tissue or organs and related them to the patient's symptoms. Thus, in the early nineteenth century medical science used the techniques of physical examination and gross anatomy to correlate clinical manifestations with the lesions discovered during postmortem examinations. It resembled sciences like botany, and the results were classifications of lesions and museum displays. A generation younger, the microscope-carrying Burdon Sanderson finished his studies with a similar idea of proof, albeit one leavened by the cellular pathology of Rudolf Virchow (1821–1902); the focus was on cells and the histological techniques necessary to study them. Physiology, the study of the normal, was still intimately related to pathology, but vivisection, which allowed the experimenter to control experimental conditions in order to isolate the function under study, had emerged as the characteristic that separated physiology from pathology. Converted to this sort of experimentalism by Claude Bernard (1813–78), Burdon Sanderson played an important role in introducing it to Britain from the German universities, where it flourished.[1]

A reason for Burdon Sanderson's success and Lydgate's failure was their very different wives. For Lydgate the fair but frivolous Rosamund Vincy, who bore him four daughters, was the final blow to his research career. Burdon Sanderson, by contrast, found the perfect partner in Ghetal Herschell (1832–1909). They traveled extensively. They learned German together. She assisted him in writing his reports. In a pattern typical of the period, Ghetal would refer to "our vaccination report" and John to "my vaccination report." They never had children, and Ghetal was John's constant companion and assistant. Ghetal was also an emotional support to the frequently morose John. In many ways their marriage was a partnership, albeit an unequal one with the focus of attention firmly fixed on John and his work.[2]

John and Ghetal shared similar political convictions, reflected by a family connection to Liberal political circles on both sides. Ghetal's brother Farrer Herschell (1837–99) was to become Lord Chancellor under Gladstone, and John's nephew Richard Burdon Haldane (1856–1928) became a prominent Liberal politician who himself was Lord Chancellor from 1912 to 1915. Like many contemporaries who were committed to social reform, Ghetal and John turned away from Evangelicalism and toward a characteristic blend of utilitarianism, positivism, and progressive liberalism.

Centering the narrative on the life of Burdon Sanderson allows me to illuminate the connections between the campaign for a scientific medicine and broader trends in Victorian culture. For one thing, Burdon Sanderson exemplified the increasing secularization of Victorian Britain. The son of devout Evangelicals who dedicated their lives almost entirely to God, he turned to the material world and devoted his life to medical practice and scientific endeavors. Like many successful mid-Victorians who had been raised as Nonconformists, he moved from dissenting religion back to Anglicanism in his adulthood. Burdon Sanderson embodied the untidy nature of the era's shift from an intellectual system rooted in religion to one based on science, bridging worlds often thought antithetical, such as the metaphysical idealist and the materialist scientific.[3]

Professionalization was another feature of Victorian Britain.[4] As the nineteenth century began, the recently industrialized nation had been dominated by capitalists, on the one hand, and the aristocracy and gentry, on the other. The forgotten middle class, however—the noncapitalist professional segment—already played a prominent role. In many ways the professions grew in reaction to the problems of industrialization, and Burdon Sanderson's position as medical officer of health is illustrative of this reaction.

He worked to undermine the entrepreneurial hegemony by utilizing public health regulation to minimize the adulteration of food and drugs and the pollution of the environment.[5]

Burdon Sanderson's career illustrated many of the tensions between the old and new ideals. In his life the clash between new professional ideals and older entrepreneurial ideals was evident. As medical officer of health, for example, he could not afford to alienate the businessmen of his sanitary district. Raised as a member of the gentry with aristocratic connections (his great-uncle had been Lord Eldon (1781–1838), the Lord Chancellor), Burdon Sanderson symbolized the alliance of the professional classes with the old landowning elite. The increasing professionalization of the era underlies this entire book. Burdon Sanderson's career also demonstrated the tensions among the emerging professions (between morphologists and physiologists, physiologists and physicians, historians and scientists, to name a few) and the tenuous nature of the alliance between the professions and the gentry.

The campaign for scientific medicine was also characteristically Victorian in its optimism.[6] The Victorians believed in progress; for them progress was evident both in the course of history and in the development of science. The ideals of scientific medicine were reinforced by the common belief that progress depended on the application of scientific methods to all the problems of life.[7] For the Victorians this fact was self-evident in a world in which, through applied science, British industry dominated the world and British might expanded the Empire. It was these attitudes that John Henry Newman (1801–90) was mocking when he denied that "education, railroad travelling, ventilation, drainage, and the arts of life, when fully carried out, serve to make a population moral and happy."[8]

The ideals of scientific medicine were also a reaction to the anxiety of the early Victorian period. In a changing and uncertain world, in which revolutions seemed commonplace and traditional institutions and ideas were under threat, many Victorians retreated to hard work and the search for solid truths. This search underpinned the belief in laboratory research. Those who, like Burdon Sanderson, espoused the ideals of the laboratory as the pinnacle of scientific medicine suggested that progress could be made against pain and suffering through careful and diligent work. This progress would result from the demonstrable truths that laboratory researchers produced.

Burdon Sanderson's career also illuminates the importance of state support to medical research in Britain. With the patronage of Sir John Simon, the first medical officer of the Privy Council, Burdon Sanderson and other young practitioners were able to conduct medical research funded by the government. The income that Burdon Sanderson received from the government eventually allowed him to give up medical practice for a research career.[9]

In the immediate background of this book are Burdon Sanderson's British contemporaries—some of whom briefly inhabit the foreground, such as Simon and Henry Wentworth Acland, Regius Professor of Medicine at Oxford. Several intersecting groups strove to popularize science at this time. Burdon Sanderson remained on the periphery of the scientific naturalism movement led by T. H. Huxley and John Tyndall.[10] The other prominent British physiologists, Michael Foster and E. A. Schäfer, were part of Huxley's group and were often impatient with what they perceived to be Burdon Sanderson's unfocused commitment to physiology. In his book on Foster, Gerald L. Geison reflected this impatience in the comment that Burdon Sanderson, "in what proved to be a life-long indecision, divided his time between physiology at University College and pathology at the newly created Brown Institution."[11] Rather than revealing indecision, Burdon Sanderson's commitment to both pathology and physiology reflected his serious desire to reform medical practice.

The failure of Burdon Sanderson to establish a research school in physiology which might seriously rival Michael Foster's at Cambridge is also revealing.[12] His failure was in part the result of institutional barriers at Oxford (which resulted in a lack of junior positions), but part of the blame was his own. Although Burdon Sanderson supported junior researchers, his leadership was stultifying. His distaste for speculation made it very difficult for young researchers to relate their own work to that of others or to feel they had contributed to the total sum of knowledge. Burdon Sanderson's own successes depended on hard, careful work and the use of complex instruments. His experimental design tended to be derived from other researchers, and all of his inventiveness was poured into modifying and developing measuring and recording devices. He was temperamentally and intellectually unable to move beyond his data to draw broad conclusions, a failing in the eyes of many colleagues.

Burdon Sanderson's Continental contemporaries stand in the deeper background. He knew such famous figures as Emil DuBois-Reymond and

Claude Bernard and became acquainted with Louis Pasteur and Robert Koch. Other less prominent men, such as Hermann Munk and Auguste Chauveau, were also important throughout his career. The experimental achievements of Continental pathologists and physiologists were a constant backdrop to Burdon Sanderson's life.[13]

From the 1870s onwards, the background was also inhabited by the antivivisectionists, a group that vehemently rejected the ideals of scientific medicine.[14] The antivivisectionists were among the first to assert that animal experiments had not improved medical treatment. Indeed, the movement to reform medicine in the image of the other emerging research sciences of the early nineteenth century, long before their results directly changed therapeutics, has long intrigued scholars. Several exploratory essays have discussed the issue of medical science and its relevance, actual and perceived, to medical practice in the late nineteenth century.[15] Most of these studies have emphasized the role of physiology as the "medical science," thus in part reflecting the aspirations of nineteenth-century would-be professional physiologists. An exception is W. F. Bynum, whose book *Science and the Practice of Medicine in the Nineteenth Century* has provided an introduction to the major issues and a broader overview of the subject for American and European medicine as a whole.

By focusing on the tortuous path of Burdon Sanderson's career, we have an opportunity to examine in detail both the meaning of scientific medicine in the Victorian context and the rationale behind the campaigns to make medicine scientific. Thus, medical beliefs in the form of theories, hypotheses, and other intellectual constructs form an important part of the narrative. But for many science was less an intellectual endeavor than a mode of material practice. In that sense this study belongs to the literature that emphasizes scientific practice over the construction of theories.[16] Incorporating most of the occupations open to a nineteenth-century medic, Burdon Sanderson's life is well fitted to a discussion of medical science, medical practice, and medical belief in the nineteenth century.

In the late Victorian period scientific medicine was seen by contemporaries as an array of medical activities based in the physical and chemical sciences, not merely medical practice rooted in experimental physiology, as was once believed. These activities included physiological research, pathological research, the use of medical technologies such as the sphygmograph, chemical analysis of food and water, and sanitarian measures. Thus, the many positions Burdon Sanderson held during his eclectic career, and the various duties he performed, can be connected to a common theme. By the standards of the day he practiced scientifically.

One of these practices was laboratory research, both pathological and physiological. The trajectory of Burdon Sanderson's career illustrates that pre-bacteriology pathological research was an important factor in the laboratory revolution in medicine, as it has been called. In addition Burdon Sanderson's career illuminates the importance of personal contact for the exchange of materials, personnel, techniques, and ideas which was necessary for success. Acceptance into the international research circles that Burdon Sanderson entered depended as much on social bonds as professional ones.

The multiplicity of scientific activities also meant that practitioners who supported scientific medicine justified their beliefs by drawing on a variety of "successes." Because science in this context extended beyond experimental physiology, contemporaries (unlike some historians) did not believe that science had nothing to offer medical practice. Of course, as John Harley Warner has pointed out, the previous generation had also considered its own medicine scientific.[17] In many ways the term *scientific medicine* was a successful attempt to garner allies; no one championed unscientific medicine. The meanings of *scientific medicine* also changed over time. The germ theory debates of the 1870s, for example, both fostered and demonstrated the growing belief in laboratory research among members of the medical community and beyond.

As this book will elaborate, it was in various disputes that the ambiguity and broadness of the term *scientific medicine* became most evident. Although most practitioners, and indeed those outside the profession, ostensibly supported scientific medicine, Burdon Sanderson's career demonstrated— through controversial activities he pursued such as vivisection, the expensive dredging of disease-causing canals, and the reform of the Oxford medical curriculum—both the lack of consensus about scientific medicine and how quickly its supporters could evaporate when faced with contentious issues.

In the public arena the most divisive conflict about scientific medicine centered around vivisection. Within the community of medical practitioners and scientists there were disagreements over the meaning of scientific medicine and the importance of laboratory research to medical practice. In the second half of the nineteenth century doctors were initially supportive of the ideals of laboratory science, until it became evident that research scientists were challenging the authority of doctors at the bedside. The beliefs of practitioners and scientists were more complex and diffuse than being simply in favor of or opposed to laboratory research. Few practitioners would deny the laboratory any role or reject all techniques derived from chemistry and physics. Similarly, few scientists would contend that clinical experience was unimportant to the practice of medicine.

The focus of controversy became the location where decisions would be made, the clinic versus the laboratory. Physicians felt that the decision-making power in medicine should continue to reside at the bedside, where they could simultaneously exert their social power and moral authority, rather than at the bench, since experiments were unlikely to impress their patients. Some physicians did not believe that the information produced through physiological or pathological experiments was relevant to practice; they continued to believe that clinical experience and perhaps clinical experimentation should guide practice. I do not want to leave the impression, therefore, that this is simply the story of the gradual triumph of Burdon Sanderson's views. As some embraced the ideals of laboratory medicine, others tried to moderate what they saw as an excessive belief in the laboratory at the expense of clinical experience.

Although the optimism of the mid-Victorian period lingered, over time it was tempered by what many practitioners saw as a lack of progress, relative to their early hopes. Britons in general were more pessimistic. Britain remained a great power, but some argued uneasily that the country was at its peak of influence and importance. Medical practitioners concurred that Britons could no longer brag that their country was uniquely advanced medically as well as militarily and industrially.

This book is organized both thematically and chronologically. Although I begin with Burdon Sanderson's birth and end with his death, I have not attempted a complete biography. Burdon Sanderson was a whirlwind of activity, and many of the events of his life find little or no place in my narrative. In the first section of the book we meet the young and Evangelical Burdon Sanderson, and see how his choice of a medical career reflected the secularization of the era. The second section follows Burdon Sanderson the clinician as he successfully makes a career for himself as a researcher in the 1860s and 1870s. In part three Burdon Sanderson's experiences illuminate the support of and hostility toward the newly established laboratory sciences in the late Victorian era.

Attracted by the prospect of doing good, Burdon Sanderson was introduced to medical science at Edinburgh. He set out with his microscope to make discoveries in the Parisian mode; dazzled by Claude Bernard, he was inspired to take up physiology. He entered practice in a milieu in which his scientific credentials were worthwhile because of the hopes generated by Rudolf Virchow's cellular pathology. Although he never joined the ranks of such illustrious predecessors, he gained a reputation and a fame of sorts as a researcher on "inflammation," "contagion," and the electrical properties of Venus's-flytraps, among other subjects.

The trajectory of Burdon Sanderson's career illuminates the contingent and dynamic nature of the transformations of the medical sciences as they played out in Victorian Britain. The secularization of British society, as the sons and daughters of Evangelicals, fatigued with the energetic religious belief of their elders, turned toward the world, played a role, as did the British empiricist tradition. Usually looking to Germany, first as a model then as a potential threat, the economic and military might of Britons gave them the confidence, desire, and resources necessary to import home the new research sciences. The utilitarian aim of relieving suffering and healing the sick was pivotal. Thus, physiology always benefited from its association with pathology, though many aimed to establish it as an entirely independent field—albeit one sustained by the fees of medical students.

There were critics, among them the antivivisectionists. Many clinicians who supported scientific medicine nevertheless deplored the hegemony of the laboratory and reached for other ideals. Indeed, the deification of the "great clinician" William Osler (who succeeded Burdon Sanderson as Regius Professor) could be read as ambivalence about the rise of experimentalism. Nonetheless, "laboratory medicine" had been firmly established, and the contours of our current medical world were evident at the turn of the twentieth century.[18]

The relationship between medicine and science has been the focus of attention for historians in part because of modern criticisms of medical practice; the hegemony of science has been blamed for practices that many today perceive as technology driven, alienating or dehumanizing. From our standpoint we may identify more with the Victorian critics than with the proponents of science as a basis for medical practice. But the Victorians who espoused the ideals of scientific medicine are too often simultaneously not credited for their own vision of medicine (or its subsequent successes) and blamed for the perceived failings of a modern medicine they never envisioned.

From Evangelical to
Medical Officer of Health

CHAPTER ONE

Choosing Medicine

> Evangelicism had cast a certain suspicion as of plague-infection
> over the few amusements which survived in the provinces.
>
> George Eliot, *Middlemarch*

John Scott Burdon Sanderson was born on 21 December 1829 in West Jesmond, Northumbria. Writing in the period after the Second Reform Bill of 1867, George Eliot (1819–80) set her novel *Middlemarch* in the two years before the First Reform Bill of 1832. In this novel she evoked the provincial England in which Sanderson and Eliot both grew up. Their lives and that of their contemporaries were defined above all by Evangelical religion.

Besides religion, a strong undercurrent in the novel and one that belonged more properly to the 1860s, was the importance of science. This was no coincidence. Eliot's longtime companion, George Henry Lewes (1817–78), was a charter member of the Physiological Society, and both were connected to the circle around the scientific naturalist T. H. Huxley (1825–95). Although she alluded to positivism and Darwinism throughout the novel, it is in her portrayal of medical practice that Eliot most directly outlined her support for science, in particular her belief that medicine needed to be reformed by a science rooted in experimental practice. Eliot was widely congratulated on the success of this aspect of the novel. Lewes wrote at the time, "Among the pleasant tidings that have reached Mrs. Lewes [Eliot] she places foremost the declaration of Sir James Paget [1814–93] that not only is there no medical detail which is erroneous but that in some respects the insight into medical life is so surprisingly deep that he could not understand how the author had not had some direct personal experience."[1]

Thus, the mid-Victorian campaign for a reformed and scientific medicine was not unique. In both Britain and America earlier generations of young medical men had sought research careers, usually without success.[2] With hindsight we can see that John Scott's birth in 1829 was fortuitous; later he was able to pursue his career in medical research at a point when, pointing to the successes of German research and building on the spirit of Evangelical religion, change was occurring. Eliot herself implicitly campaigned for such a change with *Middlemarch*. Although it was set in the early 1830s, when she wrote the novel it was not evident that material support for medical research in Britain would change so fundamentally in the 1870s. Indeed, she herself played a role in the transformation, most concretely in the founding of a studentship in research physiology at Cambridge University in memory of Lewes.

Class was another important determinant of the boy's life. Born the fourth child of Elizabeth Sanderson and Richard Burdon, the young John Scott, as he was called by those close to him had joined a family on the fringes of the landed classes. Named after his uncle John Scott Eldon, the first Lord Eldon, the boy's rank was further signified by his double surname. Elizabeth's father's will required her husband to take the name Sanderson. The family's straddling of the boundary between the landed classes and the middle class was evident in the fate of later generations. In the early twentieth century the third Lord Eldon was clearly an aristocrat; in contrast, R. B. Haldane, John Scott's nephew, was described as a quintessentially middle-class man.[3]

In the end John Scott had the family connections, education, wealth, and industry to make a career as a researcher. The trajectory of his early life illustrates the transformation of religious fervor into a zeal for science which was so characteristic of the late Victorians.

An Evangelical Childhood

Both of John Scott's parents were very pious Evangelicals. Richard, after earning a first-class Oxford degree, had began a career in law and politics. He took up two patronage appointments he had been given by his uncle, John Scott Eldon, Lord Chancellor in 1801–05 and 1806–27, and started a law practice. As Secretary of Presentations, Richard discovered that religious appointments were made with political, not religious, considerations in mind. As a result, he resigned both his political positions—he later regretted giving up the second position—and ended his political and legal career.

Elizabeth Sanderson was the only child of the late Sir James Sanderson, banker and hops merchant, and Elizabeth Skinner. The family originated in Yorkshire, and Elizabeth spent part of her childhood there with her grandmother. Upon her father's death in 1799 his wife, Lady Sanderson, became a part of the "Clapham Sect," an influential group of Evangelicals which had gravitated to the remote part of London after which they came to be named; Elizabeth was educated at their schools.[4]

Elizabeth and Richard raised their children according to the customs of the day. Most of the specifics about the Burdon Sanderson childhood home are drawn from the reminiscences of Mary Elizabeth Haldane (John Scott's sister); her point of view might be more negative than John's as she was "merely a daughter." Her memories were also colored by her later reading of Charles Dickens and Charlotte Brontë, whom Mary commended for bringing the evils of early Victorian child-rearing practices to public attention.[5]

John Scott's parents were strict disciplinarians who frequently used corporal punishment. The Sanderson children, Richard, Elizabeth, Mary, John Scott, and Jane, despite the relative wealth of the family, followed a spartan regime, which included cold baths every morning. And, of course, the children never spoke unless spoken to in the presence of adults. The Sanderson household was intensely Evangelical; the family was the center of religious education. For Evangelicals self-denial was important. Children were expected to search themselves for sin, to atone for their sins, and to improve themselves constantly. In immoderate households children might keep diaries of their small transgressions—moments of idleness, for example—and pray over them. Parents denied their children indulgences, because the children were too young to deny themselves. Denial was meant to prepare the child for the great, moral struggle that life would present, with all its associated temptations—a struggle with high stakes, for to lose meant eternal damnation.[6] The tone of the Sanderson household was evident in Mary's later recollections of a happy childhood time at the end of 1833, "only marred by the thought of my own sinfulness, which never left me." Such fears were typical of the era. Charles Kingsley's character Alton Locke expressed similar sentiments: "Believing, in obedience to my mother's assurances, and the solemn prayers of the ministers about me, that I was a child of hell, and a lost and miserable sinner, I used to have accesses of terror, and fancy that I should surely wake next morning in everlasting flames."[7]

Figure 1.1. Young Burdon Sanderson,
Bodleian Library, MS, Eng. b. 2008, item 14.

Despite the gothic nature of many of John Scott and Mary's memories, their childhoods had not been entirely unhappy. The children enjoyed being outdoors, all were avid riders, and the boys hunted as well. The Sandersons usually spent the spring and summer at their Otterburn home, where the children often walked the moors, and John Scott was a keen fisherman. Both he and Mary vividly and fondly remembered the natural surroundings of their childhood in later life and other amusements of balloon ascents and locomotives. They were a close family, and even in adulthood John Scott frequently saw his parents and siblings.

Choosing a Career

He was one of the rarer lads who early get a decided bent and
make up their minds that there is something particular in life
which they would like to do for its own sake, and not because
their fathers did it.   George Eliot, *Middlemarch*

As an Evangelical, John Scott was taught to be contemplative. His mother wrote to him when he was seven, "The Lord is at no loss for means when it is His purpose to bless, & all hearts are His to turn wheresoever He will."[8] While walking on the moors one day, the young John Scott chose a career in medicine over one in law, the traditional family career. Like many Evangelicals who could pinpoint the exact circumstances under which they became aware that God had chosen them, John Scott remembered the exact place and time on the moors where he came to this decision. He acknowledged the influence of Mortimer Glover, the medical attendant for his older brother Richard during a serious illness. Glover was interested in medical science and had written on chemistry and pathology. The two took many walks together as John Scott contemplated his future.

According to his family, John Scott had never been interested in the law; he had always been attracted by the natural sciences. Perhaps his father's failure in his own legal career dissuaded John Scott from pursuing a similar course. For a family that prided itself on its social standing, the number of acceptable professions was limited, and medicine was the only career that lent itself to a scientific training. Nonetheless, when John Scott chose medicine, for the first time he set his wishes against his parents'.

Despite their unhappiness with his decision, Richard and Elizabeth acceded to his wishes, and John Scott was sent to the University of Edinburgh, because of the renown of its medical school. Oxford, his father's alma mater, was out of the question because of the religious tests. The Sandersons' religious affiliation, like that of many of their contemporaries in this period, was ambiguous.[9] Their Evangelicism was Methodist in orientation. Mary Haldane was known in her twenties, for example, as the "handsome Methodist."[10] Certainly, in his religious writings Richard publicly distanced himself from the Church of England, most definitively in the pamphlet entitled "The Church of England identified, on the authority of her own historians chiefly, with the Second Beast, as described in The Book of Revelation, chapter XIII, verses 11–18." Thus, the family would appear to be Nonconformist or dissenting, terms applied to those who were not members of the official state church, the Church of England.

On the other hand, when John sat his Royal College examinations, his parents wrote that they would send his "baptismal certificate which appears to be necessary for passing the College of Surgeons as well as all other *Colleges* lay or ecclesiastical in these most Christian kingdoms."[11] Here they meant a Church of England baptismal certificate, which was a legal birth record. Also, the family history noted specifically that Lady Sanderson, John Scott's grandmother, had been buried in nonconsecrated ground, as a Dissenter, by implication unlike the rest of the family. While John Scott was at Edinburgh, his family moved to Devonshire for a few years, where they socialized more. Among their friends were Evangelicals within the Church of England and Dissenters. It appeared that they were among the Evangelical wing of the Church: no final break was made. The Sandersons were probably mindful that a formal break with the Church of England would not be socially acceptable in the upper-class circles in which they could travel by right and meant their children to remain within. Still, Richard's remark was likely meant ironically, given the nature of his publications about the Church of England. Although Richard had abandoned his once promising career because of religious scruples, he did not want his sons to follow the same path. Even at the height of the movement, some Evangelicals supported the established Church because of its important role in preserving social cohesion, others because they aimed to reform Britain by transforming the upper echelons of society. By the 1840s their push into the upper classes meant that many Dissenters had joined (or rejoined) the Church of England.[12]

While the Sandersons remained focused on a religious life, John Scott's own interests broadened. The competition between John's new and old lives is apparent in a letter from his father to the young university student. Richard recalled John Scott's childhood and the pleasure he had received, "when you first told me your 'experience' as expressed in these few simple lines of our old psalm:

> When in my mind I mused much
> And could no comfort find;
> Then Lord Thy goodness did me touch
> And that did ease my mind!"

This letter is the only evidence that John Scott had ever experienced a moment of conversion, the pivotal event in the Evangelical life. It was the moment when Evangelicals were filled with the comforting realization that they had been chosen; like Paul on the road to Damascus, they were among the elect. Richard probably suspected that his son's religious feelings were less intense than his own, and he hoped to remind John Scott of his earlier beliefs and to return him to the fold. John Scott's new interests emerged in the close of the letter: his father promised to pay for another glass for his microscope.[13]

The microscope symbolized the new world that opened up for John Scott at Edinburgh.[14] He had spent his first nineteen years entirely at Jesmond or Otterburn, except for one brief visit to London and a trip to Scotland. As a university student, John Scott left his family circle for the first time.

Years at Edinburgh

> He carried to his studies in London, Edinburgh, and Paris, the
> conviction that the medical profession as it might be was the
> finest in the world; presenting the most perfect interchange
> between science and art; offering the most direct alliance
> between intellectual conquest and social good.
>
> George Eliot, *Middlemarch*

John Scott Sanderson attended the University of Edinburgh for four years, from 1847 to 1851. He took the full medical curriculum.[15] The medical school was no longer at the height of its reputation, but it remained an auspicious beginning for a medical career. The medical faculty included many talented professors,[16] while the students themselves were a diverse lot. Some had experience in practice but no theoretical training; others had some training in science, and a few had received a university education. The medical curriculum had not been designed with a single organizing philosophy.

Although there were different philosophies of medicine coexisting at Edinburgh, according to Sanderson, he was molded mainly by three professors: John Hughes Bennett (1812–75), John Goodsir (1814–67), and John Hutton Balfour (1808–84). They had been students together at Edinburgh and had themselves been strongly influenced by Continental, and in particular French, medical ideas.[17] All three considered the microscope pivotal to their own work and utilized it in instruction. In many ways John Scott's microscope embodied his Edinburgh education. The three professors lectured on physiology and pathology together and related the two sciences to medical practice. They all emphasized the importance of observation and that theories (pathological or physiological) must relate to empirical evidence, rather than vice versa. The influence of the French pathological anatomy school was evident. They taught the German cell theory to their students—cells were the foundation of the unity of organic forms. Thus, other biological models were important (perhaps as valid as humans) for medical students to study. More subtly, all three professors promoted the research ideal, alongside a belief that medical science held great promise for future medical practice.

John was a serious, sober-minded student at Edinburgh.[18] He was remembered by fellow students as "tall, erect, and dignified with an intellectual countenance."[19] He worked steadily and avoided the political disorders of 1848. In a letter to his sister Jane he wrote, "The students have done their best to get up another row but have not succeeded." He explained: "Prof. Goodsir succeeded in persuading them to go home at four o'clock peaceably, though he was unable to lecture. . . . All is now quiet in Edinburgh, though as I believe far from it in London." John saved his enthusiasm for his summer work schedule. He told Jane: "I think I shall enjoy the botany class in the summer. It will at any rate be a pleasant change from present occupation, though I cannot be engaged much more agreeably than I am at present." He planned, following Goodsir's advice, "to get up at five in the summer—come & dissect from 6 till 8, and then to botany—then go home and dress and spend the day as I like, which will not be very difficult." He did admit that he was unsure about "whether I shall have resolution enough to follow this plan or not, I do not know. If I could get into the way of it, it would not I think be unpleasant."[20] Sanderson's few vacations during this period were family walking trips, during which he never lost an opportunity "to botanize."[21]

Sanderson learned to botanize from John Balfour, whom he encountered in his first year. Balfour, an Edinburgh graduate, had returned to the university as professor of botany in 1845. He was not a prolific investigator but a first-rate teacher, who incorporated botanical excursions into his courses and used the microscope for course demonstrations.[22] He explored most of Scotland, combining instruction with collecting. Sanderson found the excursions extremely enjoyable and instructive and shared his instructor's enthusiasm for botany. The two men also had had similar childhood experiences: Balfour had been raised by a strict, Presbyterian father and remained sincerely religious.

During his first year Sanderson took two courses—in general and practical anatomy—from the popular lecturer John Goodsir, who had been professor of anatomy since 1846. Goodsir was a gifted teacher and productive investigator. Rudolf Virchow had dedicated the first English translation of *Cellular Pathology* to Goodsir, "as one of the earliest and most acute observers of cell-life, both physiological and pathological . . . , as a slight testimony of his deep respect and admiration."[23] The Edinburgh School of Medicine had historically emphasized the importance of comparative anatomy. In the teachings of John Barclay (1758–1826) and Robert Knox (1791–1862), a generation before Sanderson, comparative anatomy was informed by the transcendental doctrine that there was a unity underlying the diversity of forms in organisms. Goodsir was very much a product of this previous generation.[24]

In his 1983 article about Goodsir, L. S. Jacyna outlined how Goodsir was strongly influenced by Continental, especially French, transcendentalists. Goodsir came to believe that the many structures of the animal kingdom were derived from an ultimate typical form. By 1848 he believed that the cell was this fundamental unit of all life. He wrote, "A certain form is common to all organic species, whether animal or vegetable." He continued, "The form is the globular or some modification of a sphere, and is most readily seen in the lowest scale of life, but is not confined to it, for the higher animals, on being minutely examined, exhibit the same globular character."[25]

Believing that his cell theory unified anatomical and physiological notions, Goodsir lectured on both the structure and functions of tissues. Sanderson later commended him as a teacher who "did not satisfy himself with giving a mere descriptive account of the various structures he was called on to expound, but pointed out the relations of his science to physiology, pathology, histology, etc." Goodsir tended in such discussions to prefer "biological models." He cautioned his students that, although a knowledge of the structure of the human body was essential, so too was a knowledge of the "lower" animals. In fact, the lower, or simpler, the animal the greater its physiological significance: its cells were most typical. Goodsir decided that the most basic type of living organism was the "Red Snow" (*Protoccaccus viralis*)—an entity that Robert Kaye Greville (1794–1866) had identified as a vegetable and the Harvard naturalist Louis Agassiz (1807–73) later labeled an animal. The organism thus connected animal and vegetable life, and its form was of a "rounded vesicle or cell."[26] Greville was in Edinburgh and knew Balfour during the period when John Scott attended the university.

Goodsir designed his courses to convey his vision of cells to students. He divided the class into groups, with thirty students sharing a single microscope. Before looking into the microscope, students were instructed about the object to be viewed, and the microscope was focused by the demonstrator. Goodsir was well aware that it was necessary to make analogies to known worlds in order to make the still-new, microscopic world visible.[27] John Goodsir wanted to change the practice of medicine. To his disappointment he never achieved a hospital position in which he could combine his teaching and research with clinical practice.[28] Goodsir's most lasting achievement was in inspiring medical students to take up scientific pursuits, including zoology.[29]

Sanderson did not meet John Hughes Bennett until his second year at Edinburgh. Bennett had graduated from Edinburgh in 1837 then spent two years doing clinical work in France, followed by two years doing research in Germany. He returned to Edinburgh in 1841 and gave clinical instruction and private courses in pathological histology and the use of the microscope.[30] During this period Bennett had become a proponent of Theodor Schwann's (1810–82) model of cell formation. Schwann (and Matthias Jacob Schleiden [1804–81]) had proposed that the cell was the fundamental unit of organic structure and function. Although cells could be different, they were all produced by identical processes. By the late 1840s, however, Bennett became convinced that the "molecule" rather than the cell was the true basis of tissues. For Bennett molecular investigation, using a microscope, was integral to clinical practice.[31]

Bennett was appointed professor of the Institutes of Medicine in 1848. Sanderson later recounted that Bennett had introduced to Edinburgh an "exact method of studying the characteristic phenomena of disease . . . whether by the unaided sense or with the aid of instruments." By Sanderson's account, "of the instruments then available the microscope was the most important. In 1841 that instrument, now so familiar to every student, had not found its way either into the ward or into the physician's consulting room." The reference to 1841—that is, before Bennett's official appointment—indicates that Bennett had been using the microscope in the extracurricular courses that he had taught before becoming a University of Edinburgh professor.[32]

Bennett's lectures were accompanied by microscopical demonstrations from the beginning, "for which twelve achromatic microscopes of great power, by Chevalier of Paris, would be used." He claimed not to "advance any pathological doctrine which could not be supported by anatomical or physiological data, and . . . to keep constantly in mind that the men he had to teach were being trained to be doctors." Bennett could connect his lectures to the practice of medicine, because in 1848 he was also appointed professor of clinical medicine in the infirmary. As Sanderson later said, Bennett "was thus enabled to blend together physiology, pathology and practice, as he himself expressed it, into one glorious hotch-potch!"[33]

Bennett knew that the science of medicine was ahead of the practice. In 1849 he admitted to his students that, "whilst pathology has marched forward with great swiftness, therapeutics has followed at a slower pace." He emphasized: "What we have gained by our rapid progress in the *science* of disease has not been followed up with an equally flattering success in an improved method of treatment. The science and medicine have not progressed hand in hand."[34] Nonetheless, Bennett looked to the pathology and physiology laboratory for the authority to understand and construct therapeutic theory and thus to affect practice. Like Lydgate and other Continentally educated physicians of his generation, Bennett believed that it was possible to improve the practice of medicine through scientific knowledge; for most practitioners in this era such views remained unusual.[35]

The differences between Bennett and some contemporaries were evident in the Edinburgh bloodletting controversy of the 1850s. By the mid-1850s the use of bloodletting had declined in Edinburgh, although it remained as a therapeutic option. Most physicians, such as Bennett's colleague William Pulteney Alison (1789–1859), a professor of medicine from 1842 to 1855, explained that they no longer usually recommended bloodletting because either disease itself or constitutions had changed; bloodletting was no longer indicated. In other words, clinical observation had changed practice. Bedside observation, most practitioners believed, was the appropriate authority to appeal to when supporting a therapy. In contrast, Bennett drew on a belief in experimental science when he rejected bloodletting. Bennett believed that the pathology of inflammation was most important. According to his molecular theory, inflammation was due to exudation of blood plasma—thus therapy should lead to its assimilation by assisting its transformation into tissue structures. The preferred therapy was restorative treatment, nourishment, and rest; bloodletting depleted the supply of nutrients. Bennett's medical studies

had given him a different professional identity than that of most of his contemporaries, one based on laboratory practice.[36] But, unlike his predecessor
at Edinburgh, Allen Thomson (1809–84), who had also taught Sanderson,
Bennett always remained active in medical practice and later in life "viewed
with dismay the tendency to divorce physiology from its practical relations to
the wants of the profession."[37]

At Edinburgh John Sanderson was active in extracurricular medical activities. He joined the prestigious Royal Medical Society of Edinburgh, where in
1848—foreshadowing his later research on Venus's-flytraps—he read a paper
on vegetable irritability.[38] In the 1850–51 session he was chosen by his fellow
students as one of the three presidents. In his presidential address Sanderson took on those students who found the lack of debate within the society
boring. These unhappy students compared their mundane era to Edinburgh's
glorious past, when at one time Boerhaave's adherents had fought those
of Albrecht von Haller (1707–77) and, more recently, the Cullenians had
battled the Brunonians. The references are to Hermann Boerhaave's (1668–
1738) dispute with his former student Haller and to the disagreement in
Edinburgh between William Cullen (1710–90) and John Brown (1736–88).
Sanderson's remarks demonstrated the influence of his professors—he identified himself not with Edinburgh's past but, rather, with the emerging Continental tradition of scientific medicine.[39] Sanderson stated: "Our knowledge
is now laid on a foundation more certain, more enduring than was theirs. We
are engaged, as many of us as are studying our profession in the right spirit,
in endeavouring to multiply our knowledge of actual phenomena by accu-
rate observations." Because of the "accumulation of ascertained facts," questions under discussion in the recent past were "now set at rest.[40] According to
Sanderson, the era of speculation was over, and it was no longer possible for a
Boerhaave to emerge with a central dogma. His fellow students should work
together "against those barriers which Nature is ever placing in the way of
those who would penetrate into her mysteries." He emphasized that "the one
only object which we must all have in view in every attempt which we make
for the promotion of the interests of our science, is the Discovery of Truth."[41]

John Scott Sanderson graduated with an M.D. degree from Edinburgh
in 1851; his Gold Medal thesis was "On the Metamorphosis of the Colored
Blood-Corpuscles and Their Contents in Extravasated and Stagnant
Blood."[42] The marks of his education were evident in this thesis, in which

he included clinical experience alongside pathological and physiological research. He did research on animals (e.g., frogs and pigeons) and examined human tumor cases. For both groups he performed postmortems, which included the microscopic examination of tissues. Since he could not, for the purpose of his thesis, personally collect enough cases, be they human or animal, he cited some of the extensive literature on the subject, including that of France and Germany. The thesis illustrated Sanderson's diligence and mastery of microscopy, no doubt factors in its prizewinning status.

During Sanderson's years at Edinburgh he had gravitated to a particular set of beliefs. Bennett, Goodsir, and Balfour, the professors he remembered as most influential, had many common threads running through their teaching. All three had studied medicine at Edinburgh in the 1830s. Balfour and Bennett had continued their education on the Continent—Balfour in Paris, Bennett in French and German centers. Goodsir was also strongly influenced by Continental, and in particular French, comparative anatomy. Under the circumstances Sanderson's decision to travel to Paris was neither surprising or novel, although he later regretted that his time on the Continent had not been spent in German medical centers.

Time in Paris

When Eliot alluded to Lydgate's studies in London, Paris, and Edinburgh, contemporaries understood that he was an ambitious and probably well connected young practitioner. It was no accident that the diligent young Sanderson, even a generation later, followed a similar course. The facilities of the French hospitals were well known in Britain. Well into the 1850s Paris remained the chief destination for aspirants to the upper ranks of the British medical profession.[43]

In the autumn of 1851 Sanderson began work on organic chemistry with Charles Gerhardt but soon switched to Adolf Wurtz's laboratory. Working in the laboratory daily, his main project was studying the organic compounds found in animal tissues, and he also researched the constitution of substances that could be extracted from muscle, particularly keratin.[44]

He spent his mornings in the hospitals—the Midi, Charité, St. Louis, Necker, Hotel-Dieu, and Enfants Malades. There is little record of these visits or his impressions. He joined the English Medical Society and was for a time its president. The society held weekly meetings at which students read papers on medical subjects or communicated the history of cases, followed by general discussion.[45] Sanderson's schedule was unrelenting. The events of the coup d'état of 2 December 1851, which brought Louis Napoléon to power, were an unpleasant diversion; "the terrible scenes" that Sanderson witnessed "long dwelt in his memory." Unfortunately, any letters from this period perished along with Sanderson's early papers and family memorabilia in a fire at the Pantechnicon Furniture Depository in London in 1874.[46]

Far more important to Sanderson were the physiological lectures of Claude Bernard at the Collège de France, which were the highlight of his time in Paris. "He used to say, pointing to the bust which stood above his study table," Sanderson's protégé, Francis Gotch wrote later, "that Bernard was the most inspiring teacher, the most profound scientific thinker, and the most remarkable experimental physiologist that he had ever known."[47]

The young Bernard was not yet the stately symbol of physiology he became after the publication of *An Introduction to the Study of Experimental Medicine* in 1865. But by 1851–52 Bernard had worked out some of his philosophy of physiology. His ultimate aim was to base therapy on a pathology derived from experimental physiology. Physiology and pathology were connected, but physiology was the primary science of life. In his opinion the physiological approach was superior to the rival chemical approach, championed by the German organic chemist Justus Liebig (1803–73), which emphasized reducing physiological processes to chemical transformations. Using the physiological approach, a researcher could penetrate the organism to investigate processes at their sites, while chemists remained outside the organism. Bernard did not discount the relevance of chemistry. He simply argued that reduction to chemical processes could only come from a knowledge derived from physiological experimentation; Bernard's outlook had been formed by his early nutrition research. In this early period he still coordinated experimental results with clinical and pathological findings. Bernard particularly emphasized to his students the importance of conducting experiments. He was reported to have rebuffed a student who had said he thought he knew something: "Why think when you can experiment? Exhaust experiment and then think."[48]

Bernard had been appointed as François Magendie's (1783–1855) substitute lecturer at the Collège de France in 1847, and in this capacity he gave the summer lectures. By 1850 Bernard had created his own following. Sanderson's notes—which appear to be among the earliest extant—suggest that the lectures were not yet the large formal affairs they became later in Bernard's career. An early demonstration, for example, ended with all the listeners feeling a rabbit ear for a change of temperature. The notes are somewhat sparse (they improved over time, probably along with Sanderson's comprehension of French).[49] Nonetheless, Sanderson's notes indicate the vision of physiology that Bernard conveyed.

The lectures drew heavily on Bernard's own research, although he also gave a historical overview of any subject. The course was divided into three sections: "1. Phenomena of Digestion in relation to Chemistry, Physics, 2. Phenomena of Innervation, 3. Phenomena of Reproduction, Life." Bernard devoted most of the first lecture to establishing the primacy of physiology over organic chemistry and anatomy. He outlined the utility of organic chemistry, "which has afforded the necessary instruments of Research." He then sketched the fallacies of those who depended solely on laboratory chemistry, by describing some experiments that had contradicted their conclusions. Bernard also talked about the clinical importance of distinguishing purely chemical phenomena from those in which the vital influence was necessary. In the second case, he explained, "we can increase[,] diminish or change by therapeutic means which act on the nervous system, while in the first the means which we employ must be in like manner chemical." Turning to anatomy, he again used experiments to demolish those who might attempt to make conclusions about function based solely on structure. He emphasized that "anatomic analogy can never be depended on in questions of function, in default of experimentations which is the only test."

Bernard also cautioned his audience about the difficulties of "experimentation on living beings." Animal experiments frequently produced contradictory results, but the contradictions might be explained if one carefully recorded all experimental conditions. Using a specific example, Bernard explained that the apparent contradiction between Benjamin Brodie's (1783–1862) and Magendie's experiments on the ligature of the bile duct had been due to the anatomical differences between dogs and cats.

All Bernard's lectures emphasized experiments and ended with demonstrations.[50] Typically, he would begin with his brief history of the subject and include just enough detail to make the experiments that would follow intel-

ligible. Cleverly conceived, the order of his lectures often allowed Bernard
to demonstrate his points as he spoke, integrating the visuals with his discourse. The philosophical stance he took (perhaps most suited to his audience of neophytes) was that of a naive empiricist: disputes were settled by
results, results the audience could see for itself (or should see—sometimes
the demonstrations did not work). The role of generalizing theories was not
mentioned, and students were cautioned about the dangers of hypothesizing
beyond the data.

Sanderson did not merely attend the lectures; he was also part of a small
group that experimented under Bernard's direction at the Collège de France.
Given the task of collecting pancreatic juice from a rabbit, Sanderson made
several unsuccessful attempts on some unlucky rabbits and dogs before he
learned how to insert cannulas.[51] There is no record of his attempting more
complex experiments with Bernard or of any close friendship. Their acquaintance was sufficient, however, that Bernard agreed to write Sanderson a letter
of reference and to meet with him two decades later when Sanderson was
engaged in writing a textbook of physiology.[52]

Like Sanderson's Edinburgh professors, Bernard connected physiology
and pathology to the clinic. He too was hopeful that physiology would ultimately change therapeutics. Bernard emphasized empiricism, although he
had different priorities from the Edinburgh group; in his opinion physiology
was the primary science. And Bernard's outstanding surgical skills allowed
him to carry out difficult experiments successfully—hence the popularity
of his experimental demonstrations. Bernard's lectures strongly influenced
Sanderson, who reread them twenty years later with an eye to the *Handbook for
the Physiological Laboratory,* and in January 1872 he traveled to Paris to discuss
the project with Bernard.[53]

The question arises, why were John Sanderson and so many of his contemporaries attracted to experimental medical research? Their attraction was, in
my opinion, due in large measure to earnestness—a quality that described
Sanderson throughout his life. Victorian earnestness had its origins in the
Evangelical movement and to some extent in the earlier Puritanism. Victorians prided themselves on their earnestness.[54] It was no coincidence that the
title of an Oscar Wilde play that mocked Victorian values was *The Importance
of Being Earnest.* For the Victorians to be earnest was to seek out the causes of
things, not to be content with the superficial; the implied contrast was with a

hypothetically frivolous earlier generation that had been content merely to hunt, eat fine foods, and make witty remarks. This earnestness could be seen in the religious revivals of the early century—both the Evangelical movement and the Oxford movement of the 1830s. It was also reflected in the writings of Thomas Arnold and Thomas Carlyle. Walter E. Houghton associated the creed of earnestness with the unease of the early Victorian period. The events of the 1830s in Britain reinforced these worries: the Reform Bill, Catholic Emancipation, and liberal attacks on the church meant that old institutions and formerly fixed ideas were being simultaneously questioned. William Ewart Gladstone, the future prime minister, evoked the sentiments of the era when he wrote around 1831–32: "The signs of the times are so appalling. . . . Who can look abroad and say what the morrow shall bring forth, or hope that the dove of peace may find where to rest her foot amidst the deluge that is spreading." In his opinion society was "shaken to its very foundations" by the "promulgation of new, pestilent, impracticable theories." Even "the main remaining home and hope of the cause of order throughout the world"— the United Kingdom—found itself "assailed from within as well as from without."[55] This unease had its roots in eighteenth-century rationalism and the French Revolution.

The result was the general belief, originally given its force in the Evangelical movement, that the country was in a state of crisis and could only be saved by the earnest search for saving ideas and earnest living. Evangelicals, particularly those still within the Church of England, opposed the Oxford movement, named after the group of Anglican clergy at Oxford who, beginning in the 1830s, aimed to renew the state church through a reversion to Roman doctrines and rituals—popularly caricatured in the expression "bells and smells." But both Evangelicals and those in the Oxford movement attacked nominal Christianity and the superficial life it bred. Earnestness had two facets: to be earnest intellectually was to have sincere beliefs about fundamental questions, not merely to repeat conventional notions or to resort to witticisms; to be earnest morally was to recognize that human existence was not simply the period between birth and death but a spiritual pilgrimage in which a struggle against the forces of evil was necessary.[56]

The imprint of John Scott Sanderson's early Evangelical experiences stayed with him throughout his life. In his diaries he carefully accounted for most every moment—to the point where he would record a half-hour walk with his wife, Ghetal. Resentful of time away from work, he frequently noted hours "wasted" by visitors or dilatory coworkers.[57] He could be inflexible and impatient with others who did not work as hard as he felt that they ought to, and he was greatly displeased by carelessness.

John Scott was only one of many eminent Victorians who had been raised as an Evangelical. Ford K. Brown noted the "impressive role of Victorian notables who were Evangelical in their youth." Such a list would include John Ruskin, Benjamin Jowett, George Eliot, Charles Kingsley, William Gladstone, and Leslie Stephen, among others. All maintained "the strong discipline" one would expect, but "not one of them stayed an Evangelical."[58]

Publicly, in his addresses and publications, Sanderson returned to the themes of philanthropy, hard work, and earnestness again and again. Through his writing it becomes clear (at least in his own reconstruction) that his choice of medicine as a career had not been really inexplicable; it held out the possibility of doing good. In the mid-1850s he stated, "The profession of Medicine is one which compels its members to be philanthropists." Philanthropy included the research ideal. Sanderson contended that even the selfish would be compelled by the practice of medicine "to do good." The two objects of medicine—"the prolongation of life and the alleviation of pain caused by disease"—were inherently good, but the physician's work also led "to the mitigation of every form of misery and the aiding of every class of the miserable."[59] Since Sanderson gave this address after working as a medical officer of health for a year, he could speak from his own experience about the miserable classes.

Sanderson's allegiance to scientific medicine at Edinburgh had its roots in his Evangelical childhood, but his allegiance was also a further step in breaking away from that childhood. John Scott had first asserted his independence from his parents by choosing medicine as a career. In university he grew away from the religion of his parents. His words emphasize the possibility of the miserable being saved through medicine, rather than theology (as his father had attempted). From a distance it is evident that in his twenties John Scott was struggling with his father's faith, with the intellectual burden of accepting the miracles of Christianity and the emotional burden of dwelling on the possibility of damnation. Frank Turner has suggested that, because Evangelicals put the family unit at the center of religious practice, a situation was created "whereby young persons could establish some personal psychological independence through modifying that family religion."[60]

The cooling of Sanderson's religious belief was typical of the general decline in fervent Evangelicalism at midcentury—what Boyd Hilton called "a mid-century change of mood." The 1850s and 1860s were, as John Morley put it, "the age of science, new knowledge, searching criticism, followed by multiplied doubts and shaken beliefs." One escape route was to put aside doubt and turn to practical activity. Another was to exchange the religious belief system for another. To some extent Sanderson chose both paths. In a remark equally appropriate for John Scott, Gertrude Himmelfarb described Thomas Macaulay, the historian, as a typical second-generation Evangelical, who kept the values of Evangelicism but discarded the religious trappings.[61]

Sanderson did not entirely abandon his faith during the 1850s. His movement away from the religion of his childhood was not yet clear even to himself. His attraction to scientific medicine at Edinburgh had been unproblematic, since his professors remained sincerely religious while espousing scientific ideals. There, at least, he was not made to feel that he had to choose between a belief in God and a belief in science. Sanderson's stance contradicted the contention of Houghton, who stated that Victorians had to choose between science and religion. As others have pointed out, many did not feel that they had to choose.[62] Sanderson wrote in 1853 to Ghetal Herschell, his fiancée, "You have often I am sure heard it remarked by unthinking people, that those that belong to, and especially those [that] at all devote themselves to my profession very frequently become sooner or later—materialists." He admitted that being occupied as he was "with studies in which nothing is admitted otherwise than on the direct or indirect evidence of the sense, has a strong tendency to produce a sceptical turn of mind." But Sanderson had no problem separating the domain of religion from that of science. He continued: "The great mistake which we are apt to commit is to apply the same standard to matters of faith as to matters of science. We cannot by searching find out God."[63] Sanderson never disavowed this rapprochement between science and religion. In the early 1850s he remained religious and still Evangelical in his outlook.

But the structure of the memoir of Sanderson's life symbolized his later abandonment of his parents' religion. There was no moment of conversion.

In its place was the moment, always remembered, on the moors when John Scott decided to become a physician. The lack of discussion of a moment of conversion, or its absence, signified that Sanderson did not maintain strong attachments to his Evangelical roots.[64]

The shift in Sanderson's belief system was reflected in his public addresses. In an 1868 speech to the entering class of the Middlesex Medical school he alluded briefly to the divine nature of medical practice—"Medicine is truly a God-like occupation"—but he quickly returned to earth, admitting "how few there are who, even while pursuing a divine art are animated by divine motives!" Sanderson reassured the young medical students that they did not need to seek divinity, for "the pleasure to which I refer is less elevated but more easy of attainment. It is one which every earnest upright man who has a sound mind in a sound body, may, and does, constantly enjoy." This "inexhaustible pleasure" came from the consciousness of the physician (or anyone) that "he is mentally and physically competent for his work" and "determined to use the powers God has given him for the purposes for which He has bestowed them." Sanderson exhorted the students, "Make up your minds then, that whatever other sources of enjoyment you may find in life, there is an inexhaustible source in fruitful work, and especially in professional work."[65]

The reference to God was Sanderson's only mention of the deity to the Middlesex students. At least rhetorically, he still based the qualities of being earnest, upright, and hardworking in God-given powers. But Sanderson's allusion to God was peripheral. In actuality he invoked these qualities as persuasive in their own right; for him, and many other contemporaries, they had become self-evident virtues. It could be argued that Sanderson was merely tailoring his speech to his audience, for which such values were self-evident and increasingly secular. But Sanderson's youthful religious fervor had dimmed. These statements reflected his own increasingly secular outlook. By 1883, in an address defending physiology (and pathology), Burdon Sanderson invoked similar values with no reference to God: "Every strong and true man recognizes it as his duty to act with a knowledge of the consequences of his action—to know what he is about in the fullest sense of the expression. To possess this knowledge, as I hope to be able to show you, [a physician] must be a physiologist."[66]

Although Sanderson remained a practicing Anglican, and thus continued to believe in God, as time passed religion became less and less important to his life. The pivot of his life was science and the research ideal; Burdon Sanderson chose to proselytize about science. In that sense Sanderson can be numbered among the late Victorian adherents to scientific naturalism. There were differences among its proponents, but scientific naturalism rested in general on positivist beliefs: truth was the description of phenomena; anything more was unknowable, metaphysical. The solution to the era's problems, they believed, lay in rational science. Turner has suggested that scientific naturalism arose out of a confluence of factors. First, the public health campaign, efficient farming, and later the German industrial threat made science relevant to daily life. Second, science remained intelligible to the educated layperson, and as the century progressed science was increasingly professionalized.[67]

The public face of scientific naturalism, embodied in T. H. Huxley—"Darwin's Bulldog"—was confident, arrogant, and uncompromising. Huxley campaigned for the establishment of a new scientifically trained cultural elite. Unsurprisingly, since the Evangelical influence had so marked Britain, in many ways the methodology and fervor of scientific naturalism was similar to Evangelicalism.[68] Like Huxley and his friends, Sanderson was a popularizer of science, his scientific outlook was empirical, and he wanted to professionalize science. But he was on the fringes of the core group (made up of Huxley, John Tyndall, Herbert Spencer, W. K. Clifford, and Francis Galton),[69] in part because of the medical circles in which he moved and also because of his sincere desire to change medical practice with scientific research. The Huxley group, by contrast, was not that interested in medicine; they used medicine as a vehicle to promote the emerging biological sciences.

Another difference between Sanderson and the scientific naturalists was his rejection of agnosticism. Although this difference was important, more problematic was the compromising temperament that was reflected by his ability to remain a practicing Anglican and a scientist. This reference to a compromising temperament may appear to contradict the statements made earlier about Sanderson's inflexibility, but both characterizations can be correct. He could be inflexible, often over small matters, but he tended also to avoid controversy and seek compromise.

The scientific naturalists did not aspire merely to arrive at an accommodation between science and religion; they aimed for victory. Here Sanderson was perhaps more like his father than he realized. He rejected the dogmatism of scientific naturalism as his father had rejected the dogmatism of the established church. In this sense Sanderson was much more typical of most Victorian intellectuals who, having abandoned their Evangelical roots, were uncomfortable with the arrogant agnosticism of scientific naturalism. Most preferred to retain some notion of an Absolute, whether in the language of positivism or idealism.

In late 1852 Sanderson left Paris, satisfied with his experiences and confident that he had made valuable contacts.[70] Determined to pursue a research career, he went home to Newcastle-on-Tyne. His father encouraged John Scott to consider practicing in Plymouth (where his parents had moved to avoid the northern winters). According to a friend there, "A physician was much wanted for the *county families* in this neighbourhood."[71] Over the objections of his father, John Scott rejected the intellectual isolation of a career as a provincial practitioner (who perhaps attempted to make discoveries in his spare time). From this point on his life was diverted from the path that Eliot's Dr. Lydgate followed in his doomed research career. At the end of 1852 John moved to London.

CHAPTER TWO

Medical Officer of Health

What a yet unspoken poetry there is in that very sanitary
reform! It is the great fact of the age. We shall have men arise
and write epics on it, when they have learned that "to the pure
all things are pure," and that science and usefulness contain a
divine element, even in their lowest appliances.
    Write one yourself, and call it the *Chadwickiad.*
    Why not? "Smells and the Man I sing." There's a beginning
at once.   Charles Kingsley, *Yeast: A Problem*

Yeast: A Problem, one of Kingsley's lesser novels, was first published in 1848.
Kingsley was a prominent Victorian, a well-known author and social reformer, and one of the "Christian Socialists." Much of the novel was an attack on the papacy and the Oxford movement, but it also reflected Kingsley's
strong interest in sanitarian reforms. Although satirical in tone, the quotation illustrated the contemporary identification and intersection of sanitari-
anism with Evangelical religion. Confident in their missions, the Evangelicals
aimed to purify British society figuratively, the sanitary reformers to purify
Britain literally. In such an environment a young Evangelical man like John
Burdon Sanderson, with a medical training, was likely to be attracted by sanitary reform.

In his first years in London, Burdon Sanderson felt insecure and worried
about whether he could support himself; he applied for various medical positions and started a private practice. He wanted to do scientific research, but
he knew it was unlikely that he would be paid for it. By living in London, he
hoped to keep up with scientific developments and do some research on the
side. Around this time he also began to use both his surnames in order to
avoid confusion with another London physician named Sanderson.

Figure 2.1. Ghetal Burdon Sanderson,
The Wellcome Library, London,
E. A. Sharpey-Schafer.

One of the few happy events of these anxious years was his marriage. On 9 August 1853, after a brief courtship, John married Ghetal Herschell, the elder daughter of the Rev. Ridley H. Herschell and Helen Skirving Mowbray. Ghetal and John probably met through Evangelical circles. The Rev. Herschell was a luminary within the Evangelical community; he had been born a Jew in Germany and converted to Christianity as an adult. He was later buried in nonconsecrated ground as a Dissenter.[1]

In Ghetal, John found the ideal partner for a young practitioner with scientific aspirations. She did not mind the long hours John worked away from her, and she frequently assisted him in his work. In addition, John, who had a melancholy temperament, could always depend on Ghetal for emotional support. They were both close to their respective families, and Ghetal could always be counted on to entertain John's relatives when he was busy.

In his letters to Ghetal, John expressed his worries about the future. During a visit to Edinburgh in 1854 he wrote, "Last night I lay awake late after three o'clock and felt very very wretched." He continued unhappily: "I cannot say that I wish I had not come, but certainly I have had a dismal visit and feel very far from better of it. I do not know what we shall do."[2] His medical practice remained small, but he had some success with his job applications. At the end of 1853 he had been appointed medical registrar to St. Mary's Hospital and for 1854–55 lecturer in botany. (He remained medical registrar until 1858 and became a lecturer in medical jurisprudence from 1856 to 1862.)[3]

Johns's luck turned in late 1855. "Now for the news," he wrote excitedly to Ghetal, "Mr. H. told me that Thorne had been to him to tell him that if I chose to come forward I would be certain to get the appointment of Health Officer for Paddington."[4] Mr. H. was probably Ghetal's father, the well-known pastor of a chapel in the area.[5] Burdon Sanderson referred here to the medical officer of health (MOH) positions, newly created for each sanitary district of London by an act of Parliament and later expanded to the entire country.

The MOH positions were created, it was widely believed, because of the efforts of John Simon (1816–1904), the medical officer of the Privy Council. Simon was well known as the public official who in 1855 essentially replaced Edwin Chadwick (1800–1890) when he was appointed as the first medical officer of the Privy Council. Chadwick had been removed from the General Board of Health and his position eliminated. Simon's position was newly created.[6] Politically astute and diplomatic, unlike Chadwick, Simon was both careful not to offend commercial interests and well connected. His arrival on the metropolitan scene, in 1848, as the first medical officer of health in the City of London had symbolized the triumph of medical science over engineering as the basis of sanitary reforms, in other words, the overthrow of Chadwick's views.[7] His letters of reference for his first position testified to his scientific rather than his sanitarian credentials.[8] In 1850 the *Times* asserted that Simon's most formidable asset was his scientific authority.[9] Simon's social standing also did him no harm professionally.

Similarly, Burdon Sanderson emphasized his scientific qualifications when he circulated a letter to garner support for his MOH application. He wrote that he had "abundant opportunities both in London and Paris of becoming acquainted with the most important applications of medical science to the subject of public health," and he was also "so circumstanced," as he put it, "that I am able honestly and conscientiously to offer the best of my time and labour for the fulfillment of the objects contemplated by the parochial authorities."[10]

At Edinburgh Burdon Sanderson had studied public health. Thomas S. Traill (1781–1862), professor of medical jurisprudence and medical police, taught public health as part of the required course on medical jurisprudence (along with forensic medicine). In addition, William Pulteney Alison (1790–1859), a former professor of medical jurisprudence and medical police, now a professor of medicine, continued to teach an ethos of promoting the public health: in his lectures he included the incidence of typhoid, typhus, and smallpox, reflecting his own work among the poor at the Edinburgh New Dispensary. Alison believed that practitioners could choose between two methods of promoting public health—either by framing (and facilitating the passage of) new laws or by persuading people to obey existing regulations by explaining the result of omission. He believed in persuasion and felt that public health laws infringed on the rights of individuals and should only be enacted as a last resort. Alison opposed Chadwick's assertion that fever was the cause and not the result of destitution. He resented Chadwick for narrowing the focus of public health to the removal of materials likely to become putrescent. Alison's dispute with Chadwick was part of larger schisms within Evangelical philanthropic movements.[11]

The emphasis on public health in the medical curriculum explained, to some extent, the large proportion of early public health workers in England who were Edinburgh graduates. Burdon Sanderson, however, had shown no particular interest in public health. Tellingly, on 21 January 1856, the date his application was due, he "hurried home to pack up his testimonials and take them to the Vestry Hall." The only direct connection that the writers of his biography could point to was the weak assertion that "undoubtedly when completing his studies in Edinburgh his Dispensary practice led him to the poorest parts of the city, the conditions of which . . . were not very different from those which then existed in London." In fact, he almost forgot to apply for the position, since "his thoughts had been more occupied with some scientific apparatus of which he was in search than with his immediate interests!"[12]

Nonetheless, the position must have been appealing. The job was part-time, and its salary (by 1865 it was 300 pounds per annum), combined with the pay from the other positions John was cobbling together, were enough to permit him and Ghetal to remain in London and to subsidize his research projects. This salary alone put the Burdon Sandersons firmly in the upper ranks of the middle class—under 2 percent of English and Welsh families enjoyed incomes over 300 pounds a year in 1867. Burdon Sanderson's entire earnings likely pushed him close to the upper-class category, which Harold Perkin set at 1,000 pounds a year (less than 0.5 percent of families qualified)—and this was just the beginning of his career.[13] The job was also a chance for Burdon Sanderson to contribute to the community, "do good," and an opportunity to apply his scientific knowledge, particularly his background in chemistry.

Burdon Sanderson's application succeeded, in part because of his social connections but also because of his scientific credentials. Simon had circulated to each vestry in London a list of desired qualifications for the new medical officers of health which included knowledge of pathology, vital statistics, chemistry, and natural philosophy and training in scientific and microscopic work.[14] These were quite different from a similar list that Chadwick had written in 1842—his desirable qualifications had been strictly those of engineering and actuarial statistics.[15] Simon was not directly involved in the appointment process, but it was unlikely his advice was entirely ignored. And, although the two men had not yet met, Burdon Sanderson possessed all the scientific qualifications that Simon considered necessary for the position.

The New Medical Officers of Health

There were forty-eight medical officers of health appointed under the 1855 act.[16] The group was diverse (the breakdown that follows is based only on each applicant's highest qualification): sixteen were fellows of the Royal Colleges, fifteen had university degrees, and seventeen were simply members of the Royal College of Surgeons (MRCS) and Licentiates of the Society of Apothecaries (LSA). (The last two qualifications, MRCS and LSA, were the legal minimum required of practitioners for registration.) A common career pattern was to have studied medicine both in London and Scotland: many had obtained an Edinburgh degree. The position attracted members of the medical elite: five were hospital consultants, three possessed Oxbridge degrees. There was also a coterie of those interested in medical science. Along with Burdon Sanderson, two contemporaries from Paris, Frederick Pavy and William Odling, had obtained positions. The prominent meteorologist (and father of E. Ray Lankester) Edwin Lankester was also appointed. Thus, it was not unusual to hope to use this position as a stepping stone to a research career; of the four "scientists" only Lankester was not at the beginning of his career. Three physicians with prestigious West End practices also became medical officers of health. The first appointees were recruited from all levels of the medical profession, and, because the positions were part-time, they continued to practice medicine as they had previously.[17]

The Medical Officer for Paddington

Officially, John worked for the vestry part-time, but he spent a great deal of his time doing vestry work, even with Ghetal's frequent assistance in writing reports and creating the many tables he included in them.[18] In a typical week he spent two half-days at the vestry, an afternoon conducting inspections and two hours writing reports.[19] The bureaucratic requirements alone were considerable. Burdon Sanderson reported to the vestry every fortnight and prepared quarterly and annual reports. He also became involved in the Metropolitan Association of the Medical Officers of Health, organized in 1856, which agitated for sanitarian legislation and kept the medical profession as a whole informed through the medical journals.[20]

Burdon Sanderson's reports were full of statistics—births, deaths, causes of death, repairs done, among others. The information fell into three categories: vital statistics, causes of insalubrity, and work done. He (and Ghetal) tabled the statistics and work he had done; he described the causes of insalubrity in prose. Through these reports it is possible to gain a sense of what it was like to live in Paddington at that time.[21] Details of life in Paddington exemplify why public health and the physical environment were important preoccupations of Britons and the conditions under which the new MOH's operated. In brief, the case of Paddington illustrates what it was that science was supposed to fix and, indeed, what later supporters of "scientific medicine" argued that science had fixed or at least materially improved.

Paddington was in the northwest of London, bordered on the south by Westminster, on the east by (North) Kensington, and on the west by St.

Marylebone. To the south it contained Kensington Gardens and Hyde Park. Being a mixed district, it consisted of two distinct localities: St. Mary's and St. John's. In general St. John's inhabitants were wealthier and healthier. The neighborhood was more sparsely populated; on its pleasant streets were large houses, inhabited by "the wealthier classes and their servants." A larger proportion of St. Mary's inhabitants were from the "middle and lower classes." The population of the vestry had increased steadily in the early 1850s. All the increase, however, had occurred in St. Mary's; the population in St. John's remained stable.[22] This trend remained true, up until Burdon Sanderson's departure as MOH in 1867.[23] St. John's was entirely occupied, and thus there was no room for immigration; St. Mary's, in contrast, still contained potential building sites. Newcomers were also more likely to crowd into St. Mary's if necessary. In the worst areas of Paddington houses contained on average ten inhabitants in three or four "tenements."

Although metropolitan, and mainly a residential area, Paddington, like the rest of London, had not entirely eliminated the vestiges of rural life.[24] There were over three hundred cows in the vestry, kept by twenty-seven cow keepers. Many of the cow sheds adjoined residences, and the keepers invariably lived above their charges. The district also contained numerous slaughterhouses (thirty-nine were inspected in 1856). Slaughterhouses were a particular focus of potential disease. Simon had described the dangers of slaughtering, which, he said, "loads the air with effluvia of decomposing animal matter . . . in the immediate vicinity [and] . . . along the line of drainage which conveys away its washings and fluid filth." Slaughtering was also associated with "many very offensive and noxious trades," so that around the slaughterhouse itself you would find "the concomitant and still more grievous nuisances of gut-spinning, tripe-dressing, bone-boiling, tallow-melting, painch-cooking &c." These evils in their turn would be aggravated by "defects in drainage, in water-supply, or in ventilation, or by the slovenly habits and unpunctuality of those to whom the removal of filth and offal is intrusted."[25]

Adding to the animal wastes were pets and undoubtedly many horses. Approximately twenty thousand tons of animal manure were deposited on the streets of London every year in the 1850s.[26] Between the animals being driven to the abattoir and the many wealthy households stabling horses, Paddington received its fair share.

Human waste was another problem. Paddington had sewers and a system of main drainage. But both backed up frequently, and many households' drains were of permeable materials, with the result that the sewer air filtered into the sleeping rooms. And not all households were hooked up to the sewer and drainage systems; the inhabitants of barges in the Regent's Canal used the canal as cesspools, discharging "bilge-water, night soil and other descriptions of filth" into the canal.[27] Recalcitrant drains were a common problem, particularly in slaughterhouses, where considerable amounts of blood and waste were produced. Household dustbins contributed to the mounting wastes. The canal itself and the wharves along its banks were a particular focus of smell and putrefaction. Goods and rubbish were transported on the canal. The dust heaps of Paddington were deposited on the canal's wharves, and there their contents were sorted and sifted.[28] Often manure, awaiting transport, also piled up on the wharves. All sorts of waste fell into the canal while in transit, adding to the excrement from the barge inhabitants. The result was that the canal basin was a "stagnant and fetid pool" from which "every breeze carries noxious emanations."[29] The adjacent neighborhood was the least salubrious quarter of Paddington.

The medical officer of health did not concern himself solely with the external environment. Deliberate food adulteration was a constant problem in the Victorian period. Milk was watered down. Alum was added to poor-quality flour, in order to produce whiter bread more cheaply. Another threat to the food supply was caused by insanitary conditions, particularly in the cow sheds. In 1856, 19 percent of Paddington's cows died in the three months before Sanderson's annual report. Burdon Sanderson knew the likely fate of a sick cow: "it becomes the object of the keeper without delay to dispose of it, before the disease has made sufficient progress to render the animal unmarketable. It is then conveyed to one of the slaughter houses." From the slaughterhouse it was "retailed for the most part to the indigent classes and in the lowest neighbourhoods." Before its death the milk obtained from a sick cow was also suspect. Bake houses were another source of contamination: in addition, there the workers were often housed in "conditions not only injurious to the health of the men, but disgusting in the preparation of the bread delivered to the public."[30]

Burdon Sanderson reported happily that Paddington's water was supplied by two companies that "have, since August 1855, drawn their supply of water from a point on the Thames a quarter of a mile above the village of Hampton." In addition, he noted, "the water of both Companies is now filtered on an efficient plan, besides considerable improvements which have taken place in the mode of collection and the general management of supply." But he still worried about the conditions of receptacles used to collect water for household use.[31]

By description Burdon Sanderson's job was simple. He should identify possible threats to health and eliminate them. In this era threats to the public health were believed to be easily perceived—any collection of filth was a possible source of disease. Thus, streets should be cleaned, drains should flow freely, businesses and households should be kept free of dirt. To maintain food quality, bread, water, and milk should be analyzed chemically to test their purity. The medical officer of health was not expected to inspect all premises personally but was assigned a full-time inspector. In reality, of course, it was impossible to seek out all possible health threats. Therefore, the medical officer and his inspector responded to complaints from citizens and systematically (after being given statutory authority to do so) visited slaughterhouses and bake houses. They also attempted to inspect all the poorer dwellings in the vestry. Middle- and upper-class dwellings were never examined unless a complaint were made.

The position of medical officer was also complicated by local politics. The vestry was controlled by the local oligarchy and was often a battleground between rival factions.[32] Burdon Sanderson initially profited from the factionalization of Paddington—for his appointment he had been able to muster support from both "Radicals and Conservatives." And his own solid social standing made it easier for him to negotiate with the parish officers. Despite a steady stream of national acts empowering the medical officer of health, as Dorothy Watkins noted, the social composition of the vestry officers led to a "bias against the regulation of building, municipalization of private water companies, removal of slum tenements, control of common lodging houses, underground bakehouses, slaughter houses, meat markets, demolition of insanitary property, etc."[33]

Paddington was no exception. Burdon Sanderson was astute and tactful in his dealings with the vestry, but he knew better than to offend its members. Thus, even though he knew that many expensive dwellings contained unsanitary conditions, he would not examine them unless he had received a specific complaint. No monied citizen would be harassed without clear cause;

in contrast, the poor were regularly visited. When visited, they were liable to "the enforcement of cleansing and lime-washing where necessary, and the removal of offensive accumulations." Of course, the poor did not own their own dwellings, and the owners had to pay for structural defects. Burdon Sanderson did not merely kowtow to the vestry: rather, he chose his fights carefully.

Burdon Sanderson made a calculated decision to avoid confrontation with the vestry. He attempted to stay in its good graces as much as possible. His confrontations with the local worthies illustrated that simple class alliances were not the only forces at work. Frequently, Burdon Sanderson allied himself with local residents of the "lower classes" against the commercial interests of his own class. His activities also demonstrated the opposition that even the most rudimentary improvements might encounter and not just on the grounds of expense.[34]

In part Burdon Sanderson, in concert with his colleagues, attempted to solve the insoluble. The long-standing and pervasive perception that an Englishman's home was, or should be, his castle constrained the actions of medical officers of health. Indeed, the right to privacy, when it protected bad drains or slovenliness, was antagonistic to the aims of public health reform. The hectoring tone of Burdon Sanderson's many reports underlined their educational function. In 1862 Sanderson looked forward to a time when familiarity with sanitary regulations would result in less and less need for compulsion, "for it will become apparent that the wealthiest and most influential individual must yield to the supreme law of public safety."[35] The actions of the medical officers of health were part of a slow process that fundamentally changed social attitudes. By the end of the century there were fewer social barriers to preventive health measures.[36] This story was an element in the rise of the Victorian culture of professionalism. The supreme laws to which Burdon Sanderson referred were scientific imperatives that rationalized the intrusion of sanitation workers into private homes. Scientific knowledge wielded by experts overwrote privacy.

Burdon Sanderson's duties as medical officer of health kept him busy. He inspected bake houses and slaughterhouses, which he hoped eventually to force out of residential areas altogether, he compiled statistics, and he made recommendations. He also wrote at length about the canal basin. On several occasions he had the canal drained and dredged. He encouraged the vestry to pass ordinances restricting the amount of time that sewage could remain on the docks. All of these efforts were futile, and the canal remained a putrefying focus of concern for his successor.

His background in chemistry came to bear on his examinations of milk and bread (and analysis of water quality). To check for skimming or thinning with water he tested the specific gravity of milk. The method for determining the addition of alum to bread was more complex. As Burdon Sanderson explained in a footnote: "The alumina was determined by Dr. Bernays and myself in the following manner: A weighed quantity was digested in water, and to the clear filtrate, solution of caustic potash was added in excess. The precipitate remaining after the application of gentle heat was removed by filtration. To the filtrate, after careful neutralization by hydrochloric acid, was added ammonia containing carbonate of ammonia. The gelatinous precipitate (of alumina) was washed again and tested, in order to prove beyond the possibility of doubt the presence of alumina."[37] The results of the tests indicated that much milk was adulterated but little bread was.

Despite these chemical analyses, the bulk of Burdon Sanderson's medical officer of health duties were sanitarian, in the spirit of Chadwick's urban engineering program. His discourse in the reports to the parish was the familiar one of latitudes, malarial climate, the dangers of dirt, and the importance of controlling bad odors. He recommended that owners construct drains, fix houses, and remove sources of filth. But there were important differences between Burdon Sanderson's views and those of Chadwick. Chadwick in his cry of "Dirt, disease, poverty" emphasized the interrelationship of the three factors. For Burdon Sanderson poverty was not the cause of disease; disease was caused by proximity to putrefying matter. He did not deny that there was an association between poverty and disease—after his first reports he divided his statistics not by the neighborhoods of St. Mary's and St. John's but, rather, between "houses inhabited by the poor" versus "the rest of the Parish." He also gave a figure for "healthy districts" to use as a comparison for the entire parish. In general all indices of health (particularly infant mortality) were worse for the houses of the poor than for the model healthy district. The rest of the Parish, in contrast, tended to match the figures for a healthy district. But Burdon Sanderson believed that poverty contributed to ill health; the poor were less healthy because they tended to live in or near sources of filth.

Burdon Sanderson (and Simon, among others) held different views from Chadwick about how dirt threatened health. Chadwick believed that disease was caused by the toxic, inorganic products of decay.[38] The alternative theory was Justus von Liebig's view that decay was a necessary process in which organisms were dissolved into inorganic nutrients to feed plants. Putrefaction was the first step in organic decay, but it could also lead to disease. If an organism with a weak vital force encountered decaying or putrefying matter—a "ferment"—the organism could be infected and start to putrefy in its turn. Thus, for Liebig many disease processes were at root particular kinds of fermentation.[39] Burdon Sanderson had been taught chemistry at Edinburgh by William Gregory, a former student of Liebig, and his views of putrefaction were clearly influenced by Liebig's.[40] He attributed the malevolence of the canal, for example, to the fact that "the water of the Canal Basin contains a large quantity of animal matter, partly suspended, partly dissolved; this consisting either of living animalculae or semi-putrescent debris." The offending agents were the putrefying debris and adjacent organisms, not the products of putrefaction. Whether one ascribed to Chadwick's or Liebig's theory, there was one clear mode of action: the canal needed to be cleaned out.

Attracting John Simon's Notice

Burdon Sanderson brought himself to Simon's attention by sending him one of his first reports as a medical officer of health. Simon replied cordially, commented on the cow mortality figures, and asked for more information.[41] Like Burdon Sanderson, Simon was a physician interested in research who wrote passionately and at length that scientific research must be the basis of medical practice.[42] In 1842 he asserted that medical students usually lacked sufficient preliminary education. Because of this lack, the students were "bewildered with facts they cannot group, with analogies they cannot apprehend, with reasonings they cannot follow." Simon's solution was to require that students be "instructed in the elementary sciences of nature," which he defined as "zoology, botany, and physics," before their medical studies.[43]

Simon also believed that physiology was essential to pathology. He wrote, "Let me impress upon you . . . that any supposed science of disease must of necessity be crude or fictitious, unless it be a direct deduction from the knowledge of health."[44] Friendship and mutual respect developed between the two men.[45] As we shall see in the following chapter, Burdon Sanderson's hopes to develop a research career matched Simon's aspirations to expand his department's mandate to include medical research.

Simon's patronage began with a few small commissions for Burdon Sanderson, who in 1857 wrote to tell Simon that he had decided to investigate diphtheria; Simon then asked him to undertake a diphtheria inquiry for his department.[46] In 1860 Burdon Sanderson was one of many chosen by Simon to be an inspector for the Medical Department of the Privy Council; his first duty was to take part in the inspection of vaccination practices in the country.[47] These appointments brought Burdon Sanderson into more frequent contact with Simon and were a source of further income (he was paid three guineas a day while on government business). Burdon Sanderson traveled outside London in order to carry out inspections, and, although the duties were not onerous, they could be disheartening.

Rejecting the Provinces

John's letters to Ghetal, written while traveling on Privy Council Business at this time (ca. 1860), illustrate that John's reasons for remaining in London were not purely professional. He had a positive dislike, bordering on disdain, of provincial life. On one occasion he wrote: "Today has been rather a vexing day. Only one medical man out of 4 appeared at the place of meeting. The Clerk was to blame. He was very little disposed to be obliging—put me in an uncomfortable room. As I was holding my interview with him, a wasp crept up his leg & stung him. This served him right, & happily I had finished my business."[48] In another letter John described an evening party: "There was a large majority of ladies, only 2 or 3 gentlemen—very characteristic altogether of such a party as you might expect in a little town where there is no trade—no gaiety—no nothing. The ladies amused themselves the whole evening from 7 P.M. until 11 P.M. with spelling out words—the gentlemen with looking on silently." The evening did not improve during supper, John complained, "the conversation being—How did you like the concert—or the lecture at the Institute—or when will the railway be completed. These topics being, as I was afterwards informed the *staple* of Lancaster conversation."[49]

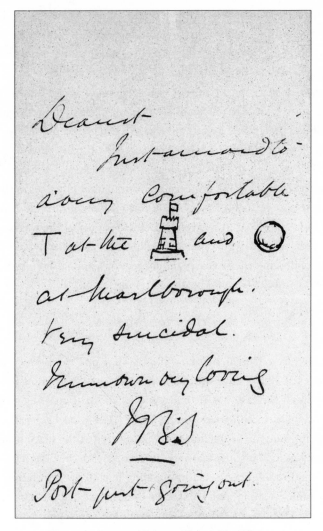

Figure 2.2. Letter of John to Ghetal, The Library,
University College London, MS ADD 179/97.

Burdon Sanderson's opinion of provincial life was an explicit rejection of his childhood milieu. It was also an implicit rejection of Evangelicalism, which, as George Eliot vividly noted, had played a key role in eliminating provincial amusements.

Besides the inquiries for Simon, Burdon Sanderson was also involved with numerous other activities, some related, others not. He examined the ventilation system at St. Mary's Hospital. He deplored the lack of scientific research on the subject of ventilation. During a journey to Wales in 1855 he took air samples in valleys and on mountains, in order to compare them with the air of towns, both healthy and unhealthy. In 1857 he participated in a Whitehall Commission on Adulteration—to study the illegal practice of tampering with food for commercial gain, for example, the watering down of milk before sale.[50] In late 1859 he became the assistant physician to the Brompton Hospital for Consumption. He had joined the Society of Observation, which focused on pathology, immediately upon his arrival in London. An extract from law 3 read, "Each ordinary member shall be expected to read the records of two cases, or to read the records of one case, and to give ten answers to questions for the special phenomena of disease, during the year."[51] And he became a member of many other medical societies, including the Pathological, the Clinical and the Medical and Chirurgical.[52]

He continued to do research in whatever spare time he could snatch. There are numerous references to his purchase of apparatus, attendance at scientific meetings, and research activities. He wrote in the late 1850s: "I have made great progress with my second wife [probably a new piece of apparatus]. She is becoming more obedient & docile and is losing her habit of always answering wrong and being a long time about it."[53] The good-humored interjection was Ghetal's own later note. Clearly, she was not insulted by being likened to a piece of equipment, doubtless recognizing John's near obsession with instruments of all sorts. During this time his chief research interest was the mechanical and chemical processes of respiration.[54]

As the 1860s began, Burdon Sanderson had reason to be pleased. Through his own diligence he had made a career in London, and he no longer needed to fear a future as a provincial practitioner. Professor Hughes Bennett had cautioned him when he had moved to London "not to be led away by the seductions of life around him to leave science for money-making practice,"[55] and he had not succumbed. He maintained an active research career. The future looked hopeful, but it was not enough, and John was frequently morose; he wrote to Ghetal: "I feel more isolated on this journey than I have ever done before—why I am sure I don't know. . . . But I do wish for home settlement so much and how far it is still from being realized. Indeed this year I don't think we shall have any quiet at all, especially if Murchison gives up Middlesex and I have to go in for it. O that will be horrid."[56]

Science and the Health of Britons

Despite the somewhat eclectic nature of many of his activities, Burdon Sanderson integrated them in his mind. His saw his clinical work as an opportunity to observe pathological processes at work—processes he hoped to prevent, of course, through his work as a medical officer of health. In his reports to the vestry he often included information from his latest research project. The reports also drew heavily on the research of others; he was immersed in the relevant scientific literature.

Burdon Sanderson's faith in the utility of science colored his experiences as a medical officer of health. The reader is left with the impression that, although there was still much work to be done—remember the canal—the health of Paddington had been improved by his activities. Not that he ever personally took credit for the improvements; they were due to the general movement for public health and credit could be assigned to many. Running through his reports was an optimistic tone belied by his own data.

In fact, there was not much evidence that the overall health of Paddington residents had improved during his tenure as MOH, as figure 2.3 illustrates. The numbers must be considered estimations; for example, the population figures after 1861 were based on the 1861 census of the district, and the figure for 1856 was Burdon Sanderson's own estimate.[57] Burdon Sanderson organized his mortality tables to highlight the numbers of deaths of children under five, since, as he noted "the mortality of infants is a much more sensitive index of the general health, than that of persons in the more advanced periods of life."[58] In addition, as he admitted in 1862, "the vitality of the population falls far short indeed of that which prevails in rural districts." "But," he concluded, "better results may be hoped for, for many salutary changes are in progress, and some are on the eve of accomplishment."[59] Burdon Sanderson's statistics, and to some extent his optimism, are supported by Anne Hardy's study of the health of Londoners and the effects of local preventive medicine. As Burdon Sanderson experienced in Paddington, the health of the population did not radically improve in the 1850s and 1860s, but 1870 was to be a turning point, and, according to Hardy, it was the efforts of the medical officers of health "which led to the eventual eradication of the epidemic streets."[60]

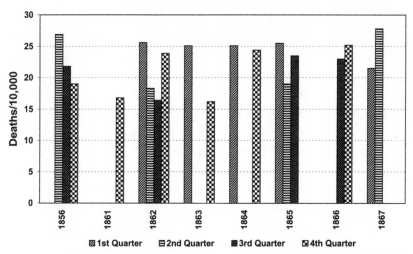

Figure 2.3. Mortality of children under five, Paddington.

Perhaps Burdon Sanderson perceived some unquantifiable, yet promising, signs around him as he carried out his tasks as MOH. His energy and optimism were typical of the mid-Victorian spirit, which believed that solid hard work would eventually be rewarded by results. This "Victorian frame of mind" played an important role in spurring on those who, like the London MOHs, worked for decades without many apparent rewards.

This is not to say that there had been no successes that an MOH could point to before the 1870s. The clearest example of the role science had played in public health was water. Burdon Sanderson explained that, through the authority of Simon, it had been "determined in the most satisfactory manner" that a population's health depended on "water containing a large proportion of organic purity." Without mentioning John Snow (1813–58), the investigator, Burdon Sanderson went on to relate the familiar story. One street had received its water from two different companies: the Lambeth and the Southwark and Vauxhall. Those who had received the fouled Southwark and Vauxhall water (which was from a Thames' source downstream from that of the Lambeth) died at over three times the rate of the Lambeth customers during the 1854 cholera epidemic. The result was that Paddington's water (among other localities) was now drawn upstream and was relatively safe.[61]

The actual versus perceived health of Britons in this period remains a contentious issue. How do we measure health? And whose health was the measure of the country's? As Paddington illustrated, the wealthy were still more likely to be healthy than the poor. There were also geographical variations — the countryside was healthier for its residents than the cities for theirs. Nonetheless, there appears to be a consensus among historians and demographers that Britons were healthier than the citizens of other industrialized nations in the late nineteenth century.[62] Of course, this relative healthiness could be explained by factors other than hygiene, in particular by improved nutrition. Christopher Hamlin noted that from the midcentury onward "British sanitary systems became the universal mark of adequate public provision for health."[63] Rightly or wrongly, contemporaries believed that the public health movement should be credited in general for the increasing health of Britons.

For Burdon Sanderson, British sanitary reforms were based on scientific principles. After visiting St. Petersburg, he wrote, rather smugly, "that any epidemic disease, and especially typhus fever, should be more fatal in St. Petersburg than in London can scarcely be a matter of surprise to us." He continued, "Indeed, if it were otherwise, we should be compelled to admit that the fundamental notions on which our efforts for sanitary improvement are founded, are erroneous, and that the sacrifices we have made to render our capital the most healthy of modern towns, have been useless."[64] Burdon Sanderson, like many Victorians, was confident that public health measures had made a difference in London.

PART II

Making a Career in Medical Research

CHAPTER THREE

Before the Germ Theory

The Cattle Plague of 1865–1866
and the State Support of Pathology

Against a disease which is highly contagious, undiscoverable at a
certain stage, and too widely diffused for an army of inspectors
to cope with it, there is clearly but one remedy which would be
certainly and absolutely effectual. That remedy is, to prohibit
everywhere for a limited time any movement of cattle from one
place to another. Enforce this, and, within a time which cannot
last very long, the disease is at an end. It must stand still, and it
must starve for want of nutriment. This great sacrifice would
certainly eradicate the evil; we cannot say so of any sacrifice less
than this.   *Report of Royal Commission into the Origin and*
Nature of the Cattle Plague, 1866

John Simon was pivotal to the creation of a community of research pathologists in Britain. Going against the laissez-faire spirit of the period, he
used his position to assemble government funds with which to support a
growing number of medical researchers.[1] Demonstrating his mastery of bureaucratic politics, Simon used several strategies to get these funds. Frequently, he would argue that a very important project required one small
grant to accomplish. If successful, as the project neared completion Simon
would contend that the circumstances discovered by his investigators had
demonstrated that the project needed to become permanent. In addition,
whenever possible, Simon dealt with the Treasury directly, sidestepping Parliament. He would maintain that Parliament had already granted money for
a project that at first sight appeared entirely novel to the Treasury; he thereby
eliminated the need to get specific parliamentary sanction.

Despite the British government's lack of interest in direct funding of scientific research, the budget for this research increased throughout the second half of the nineteenth century. Peter Alter charted the budget devoted to scientific investigations, which included subsidies to scientific societies; funds for scientific institutes; expeditions; and academies, public libraries, and music academies. Although Alter noted that "it is very difficult to gain any general overview of total state expenditure on science and research in Britain between 1850 and 1900," he estimated that total expenditure went from £120,720 (1869–79) to £215,510 (1879–89) to £267,769 (1889–99) to £568,609 (1899–1909).[2]

Simon's mandate included the investigation of diseases and their causations, and an outline of the research he commissioned provides insight into the sorts of activities which were subsumed under the rubric of scientific investigations. Initially, Simon fulfilled this requirement by sending out researchers to record incidences of diseases, such as diphtheria, in various localities; that is, he sponsored epidemiological research. He developed the methodology that came to be used for all disease occurrences investigated. The researcher was to note: (1) the general features of the district; (2) the duration as well as local and personal causes of the diseases; (3) other related sickness; (4) the communicability, symptomatology, and treatment of the disease.[3]

Simon also persuaded Parliament to fund a systematic examination of the practice of vaccination for smallpox in Britain: inspectors went county by county and examined the practices and qualifications of the local inspectors of vaccination and the quality of the lymph. His argument depended in part on an appeal to British pride; he emphasized how disgraceful it was that in Edward Jenner's (1749–1823) own country vaccination practices were often deficient. Jenner had popularized the technique of immunizing against smallpox through inoculation with cowpox lymph (i.e., vaccination) in 1798. The vaccination inquiry was a large task and required a substantial grant: Simon requested four thousand pounds for 1863–64 to finish the project. In his familiar pattern he managed to create an ongoing project out of what he had originally argued would be a one-time expense. After the inspection was completed, he concluded that "there has grown up a system of public vaccination so full of errors and irregularities" that the inspection should become permanent and two thousand pounds a year should be devoted to the purpose. The Treasury agreed but attempted (unsuccessfully) to move the necessary money from Simon's budget to the new National Vaccine Establishment. Simon staved off any attempt, no matter how small, to lower his departmental budget.[4]

Burdon Sanderson was a member of Simon's team. It was Simon who directed the studies and divided up the localities among different practitioners. In concert with other doctors, Burdon Sanderson researched and wrote parts of the diphtheria epidemic study and the vaccination inquiry. Although Burdon Sanderson was well paid for his work on vaccination, in 1863 Ghetal explained to her sister-in-law Jane Burdon Sanderson, "We have at last finished our vaccination report & are heartily glad to have seen the last of it." Ghetal continued: "John has taken to more congenial employments the particulars of which I do not enquire into. I only hear that certain rabbits & guinea pigs have been purchased."[5] In 1865 Burdon Sanderson traveled to Dantzig, at John Simon's request, to investigate an epidemic of cerebro-spinal meningitis for the Privy Council.

Simon's organizational style was evident in his detailed memorandum to Dr. Whitley (one of Burdon Sanderson's colleagues) before a trip to investigate an epidemic in St. Petersburg. The purpose of the trip, he wrote, "is that Her Majesty's Government may have his report, founded on personal observations, as to the *nature* of the fever or fevers now epidemically prevailing there." The Government "especially wishes to learn if any fever now existing in St. Petersburg is of a kind not habitual to the United Kingdom." Even if the fever (or fevers) was "not different in kind," Whitley should decide if it "is modified in any important particular from forms with which English practitioners are familiar." He was immediately to "satisfy himself as to the essential nature of the prevailing fever or fevers" and telegraph the information to London, and he should then "more leisurely" determine its progresses and causes, telegraphing or writing his findings regularly.[6]

The fruits of Simon's system were evident in Burdon Sanderson's report of his trip to the Lower Vistula. Burdon Sanderson's mission was to determine if the epidemic disease there was related to the concurrent epidemic in St. Petersburg and whether it was likely to invade Britain. The report followed the standard format Simon had set out for all epidemiological reports. Burdon Sanderson first outlined the weather conditions that had contributed to the disease, the long and severe winter and the malarial conditions that prevailed. Then he discussed the possible courses of the disease, including postmortem observations. In his summation he highlighted the poor ventilation and drainage in the most affected regions. Throughout his report this region was compared implicitly and explicitly to Britain and found wanting: "in Germany the evil effects of overcrowding are very much aggravated by the want of ventilation," and "the climate of the neighbourhood of Dantzic differs from that of London in being less equable." The conclusion was that Britons should be safe from this epidemic. Burdon Sanderson emphasized that the variables listed were the only ones implicated in disease causation. He wrote decisively, "No facts were met with in the course of the inquiry which afforded ground for believing epidemic meningitis was capable of being communicated by personal intercourse."[7]

Shortly after his return from Dantzig, Burdon Sanderson was asked by the Cattle Plague Royal Commissioners to study the nature of the disease. Initially pleased, he wrote to his sister, "All this will be very labourious but I need not tell Jenny that it is not the sort of work that is irksome to me, so that I do not grumble at it as I used to do about the vaccination work." But by the inquiry's end he was "sorry that it has occupied so much more time than I intended; for, for the last 5 months it has taken up all my available leisure."[8]

The Cattle Plague

In June 1865 the cattle plague had been reported in Islington, a district of London. By July it had spread throughout London cattle and into Suffolk, Shropshire, Norfolk, and Scotland. The disease, also known as Rinderpest, spread rapidly. In October there were over 11,000 cases, with infected cattle in twenty-nine English, two Welsh, and sixteen Scottish counties. By January over 120,000 animals were known to be infected, and there were no signs of abatement. The disease had also spread to the Netherlands, but the French and the Belgians avoided the contamination of their cattle by placing an embargo on all British livestock. Although the epidemic did not yet directly threaten domestic food supplies—the estimated total head of horned cattle in Great Britain at the time was seven million—the swiftness of the spread of the disease and the damage to British trade led to the appointment of a Royal Commission, which reported to Parliament three times in 1865 and 1866.[9]

The Royal Commission on the Cattle Plague marked an important change in attitudes toward pathological research in Britain. The etiology of this usually fatal disease was obscure; it was known to be contagious and to incubate within otherwise healthy animals for about a week. Faced with an old but little understood disease, the commissioners hired several researchers, including Burdon Sanderson, to investigate the cattle plague. This decision demonstrated that, at least in political and scientific circles, there was an expectation that laboratory research could produce information that would change practices for the better. The decision to devote government funds to fundamental medical research, long an aim of Simon, also became an important precedent. Afterward Simon was able to fund research into human diseases. The testimony to the commissioners and the research itself, later used somewhat dubiously to prove that Britain had been in the vanguard of germ theory research, exemplified the complex medical beliefs of the pre–germ theory era.

The reports of the Royal Commission captured the prevailing beliefs about contagious diseases in the period immediately preceding the new articulation of what became known as "the germ theory." The work of the Royal Commission took place before the research of Louis Pasteur (1822–95) and later Robert Koch (1843–1910) had become well known within the community of educated medical men discussed here.[10] The reports illustrate how the complexity and ambiguity of contemporary beliefs meant that, even as Pasteur's germ theory became known in Britain, most famously through Joseph Lister's (1827–1912) efforts, it did not take the British scientific community by storm but, instead, was received as yet another layer of explanation to be grafted on to the existing explanations.

Initially, Burdon Sanderson was disappointed that his cattle plague research occupied most of his time from November 1865 to May 1866. But in the end this project was a pivotal moment in his career. That he was chosen to participate demonstrated his acceptance within that British circle of men who, looking to developments on the Continent, wanted to foster pathological research in Britain. Second, the success of the project meant that he received ongoing research contracts from Simon, which became an important source of Burdon Sanderson's income. He added the cattle plague project to an already full schedule; in this period he juggled his medical officer of health job, various clinical positions in London, a small private practice, and as much laboratory work as he could fit in. After the cattle plague investigations Burdon Sanderson devoted more and more of his time to research and less time to medical practice.

The Royal Commission on the Cattle Plague Epidemic of 1865–66

The Royal Commission on the Cattle Plague began on 29 September 1865. The twelve members of the commission can be divided roughly into two groups: seven of them were scientific or medical men; the remainder represented the landed interests that still dominated Parliament. In the first category were Henry Bence Jones and Richard Quain, successful physicians (Bence Jones was also noted for his chemical researches.); Thomas Wormald, a surgeon from St. Bartholomew's Hospital; Edmund Alexander Parkes, a medic prominent in the field of public health; Charles Spooner, a veterinary surgeon, who had become president of the Royal College of Veterinary Surgeons in 1858; Robert Ceely, an authority on the natural history of vaccination; and Lyon Playfair, a chemist and Liberal Member of Parliament, who often represented medical interests in Parliament. The other commissioners were John Poyntz Spencer (fifth Earl Spencer), Viscount Cranborne (later Prime Minister, as Marquis of Salisbury), Robert Lowe, Clare Sewell Read, and John Robinson M'Clean.[11]

The hand of Simon, the highest-ranking public health official in Britain, was evident both in the choice of these commissioners and in the direction they took in their inquiries. Simon was well acquainted with all of the scientific group and a friend of many of the medical men. Furthermore, Lowe, who led the public debates and was very interested in public health generally, was a great friend and admirer of Simon. There is little doubt that his opinions were strongly influenced by Simon.[12]

During the course of the epidemic the commission made three reports to Parliament in 1865–66. The commissioners presented their first report on 31 October 1865. It included the minutes of testimony, a summary of their conclusions, and a supplement detailing sanitary recommendations. The disease continued to spread, and the commissioners submitted their second report on 5 February 1866. Here they reiterated their earlier recommendations and explained that they had initiated a series of scientific investigations into the cattle plague. The results of these scientific investigations made up the bulk of the third report, which was presented on 1 May 1866.

Although the commission was established to investigate an animal epidemic, from its inception the evidence collected and the testimony given to the royal commissioners connected the cattle plague to human diseases, most commonly smallpox, and the progress of the epidemic as well as the scientific investigations were extensively covered in the medical press. A Dr. Crisp reported in the *Transactions of the Pathological Society* that "the poison of the 'cattle-plague' (with the exception of a papular or vesicular eruption on the skin) produces no injurious effects when introduced into the human system." This conclusion was derived from Dr. Crisp's own consumption of parts of diseased oxen. The commission's reports and the press reaction also illustrated how contemporary views of contagious diseases (those caused by the spread of contagions) overlapped with beliefs about noncontagious diseases that were believed to be caused by environmental conditions such as filth. It was believed that the weather could hinder or assist the dispersal of the contagion.[13]

The cattle plague commissioners were preoccupied with establishing exactly the foreign—presumed Russian—origin of the disease, although, as Simon pointed out, since the disease was now firmly established in Britain, its origins had become a rather moot point. This question is another instance in which the germ theory issue overlapped with the debate over spontaneous generation. Addressing this question, Simon testified that the zymotic diseases were unlikely to originate through "spontaneous generation," although historically there must have been a first case of any given disease. The evidence was against such creations being common, since there were several examples of isolated regions that remained free of a specific disease.[14] The commissioners accepted from the outset the expert testimony that the disease was contagious, writing in their first report "that the disease in question is contagious, that the contagion is extraordinarily swift and subtle, and that it is most destructive in its effects, there can be no doubt whatever." They did not find it necessary to define what *contagious* meant. But, as the language in the first report illustrates, to be contagious was to be spread by the movement (however mysterious) of a physical substance, the "contagion." The cattle plague contagions were also described as "the seeds of the disease" and "the particles of the poison." In the case of the cattle plague the commissioners, following the expert testimony they had heard, believed that the sources of these unknown contagions were "the excretions from the diseased animal."[15]

As historian Margaret Pelling has pointed out, it is not possible to characterize any person or theory as simply contagionist or anticontagionist. There were never two poles but, rather, a continuum of contagiousness with smallpox (definitely contagious) at one end and intermittent fever (definitely not contagious) at the other. In between were many intermediate, and thus controversial, diseases such as yellow fever, typhus, and cholera. The middle of the nineteenth century was not, as once was believed, the height of anticontagionism but a period of general adherence to a newer concept of contagion. Pelling challenged E. H. Ackerknecht's classic account, which connected anticontagionists to the ideology of free trade; in his formulation anticontagionists opposed contagious theories of disease because they resulted in restraints on trade. The cattle plague debate did not pit anticontagionist free traders against contagionist quarantiners. All twelve commissioners agreed that in theory a quarantine would be effective against the disease (as the French and Belgian examples had amply demonstrated).[16]

In the first report the majority's sole remedy was "to prohibit everywhere for a limited time any movement of cattle from one place to another."[17] Four dissenting members—Earl Spencer, Viscount Cranborne, Mr. Read, and Dr. Bence Jones—presented themselves as realists, arguing that a tight quarantine would be ineffective because it would encourage evasion or, worse, social unrest. They asserted that it was not "practicable" to stop the movement of cattle because an effective quarantine would result in a sudden rise in the price of meat in towns. The fear was that there would be demonstrations or even riots carried out by those suddenly unable to afford meat. Instead, the dissenters recommended the less stringent measures in the report and added that animals should be licensed for movement. A fifth dissenter, Mr. M'Clean, argued separately that any extraordinary restrictions on the movement of cattle were unwarranted and that the powers already vested in the Privy Council were sufficient to prevent the spread of the outbreak.

Since none of those members who dissented from the majority report were, with the exception of Bence Jones, scientifically trained, one could see the cattle plague debate as a battle between the new scientific experts and the landed interests, who had long dominated Parliament. But this, too, would be oversimplistic, since all of the commission members accepted the scientific consensus that the cattle plague was contagious. The disagreement centered on the practicality of the measures the majority had recommended. John

Fisher suggested that perhaps Bence Jones voted with the landed interests because of his own family connections to the gentry.[18] The commissioners were very sensitive to the risk of social turmoil, in part because the commission and the epidemic were occurring in the shadow of debates over the Second Reform Bill to enlarge the British franchise, which was passed by Disraeli's government in 1867.[19]

In his testimony, which was typical of the pages of expert medical and veterinary testimony which the commission recorded, Simon closely associated the cattle plague with human epidemic diseases. Cattle plague was a specific fever, one of "several zymotic diseases," it was "exceedingly contagious," and, like similar human diseases (smallpox, scarlet fever, typhus fever, typhoid fever), it had "no *cure*." He explained that each of these diseases had its own contagium. "The contagium of small-pox will breed small-pox, . . . the contagium of typhus will breed typhus," he stated, making the point that "the process is as regular as that by which dog breeds dog and cat cat" and "as exclusive as that by which dog never breeds cat, nor cat dog." It was only dogmatists who would conclude either "that there must have been an overlooked inlet for contagium" or "that there must have been in the patient's body an independent origination of the specific chemical change."[20]

On the whole Simon held the contagionist line in his testimony. In his opinion cattle plague—a contagious disease—was not caused by filth. He stated unequivocally, "Epidemics, after periods of absence, return without any ostensible deterioration in common sanitary circumstances."[21] The commissioners were not entirely convinced; they kept returning to the fact that the epidemic had originated in London, an area where the cow sheds were most unclean. Later they brought in several London medical officers of health to testify about the undesirable conditions in metropolitan sheds.

Even for Simon the issue was not so simple; he distinguished diseases such as scarlet fever, smallpox, and scarlatina from diseases like the epidemic diarrhoeal diseases and typhoid fever, which *were* the result of "local uncleanliness."[22] To further confuse the issue, Simon (and the commissioners) believed that "external" or "atmospheric conditions" affected the *rate* of transmission of cattle plague.

Simon's testimony encapsulated the beliefs of educated medical men interested in theories of disease causation. Their views were rooted in the theories of Liebig, the German chemist who in the early 1840s had explained disease processes in terms of fermentation.[23] The testimony also demonstrated the influence of William Farr (1807–83), statistician at the registrar general's office. Farr had coined the term *zymotic diseases* to encapsulate the epidemic, endemic, and contagious diseases; *zymotic* was derived from the Greek term meaning "causing fermentation." This category was so well known that the commissioners referred to it merely as "the zymotic class, as formed by Dr. Farr."[24] As Christopher Hamlin wrote, "the essence of the concept of zymotic disease was that disease was a spreading internal rot, that it came from an external rot, and that it could be transferred to others."[25]

Formerly contagious and epidemic diseases had been rigidly separate categories. Contagious diseases, like smallpox, were transmitted by a material poison that produced characteristic symptoms; they tended to attack a person once in a lifetime. Epidemic diseases, like fever or cholera, arose from local causes; they could attack the same person many times and varied in symptoms from patient to patient. Under Farr's scheme both contagious and epidemic diseases were zymotic; that is, they were caused by the introduction of an animal poison specific for the disease into a susceptible person's blood. Farr had in mind the action of chemical poisons or inoculated contagions. The pathological effects of such poisons were well known—animal poisons were only different because they could reproduce themselves in the blood.[26] Farr also believed that environmental conditions and especially atmospheric changes were important contributing factors in the spread of zymotic diseases.

At the close of Simon's testimony he was asked by Dr. Quain, an acquaintance and fellow proponent of pathological research, whether the commission should institute inquiries into the pathology and treatment of the disease. Simon, who had followed recent research in Europe, believed that, in order for further progress to be made in the prevention of disease, laboratory research into the causes of disease should be supported by the Treasury. Not surprisingly, he replied to Quain, "I think it would be much regretted if so excellent an opportunity as offers itself should be lost without a scientific investigation of the disease."[27]

Although Simon's influence was pivotal, the commissioners' decision to support research on the cattle plague demonstrated that a small but influential group in Britain already supported the investigation of disease processes in the laboratory. This attitude was evident in the reports that circulated in the medical and popular press of other research that had been carried out even before the royal commission began. The commissioners' decision also reflected their optimistic belief that the cattle plague could be easily prevented. In particular they anticipated that cows could be protected from the disease in the same way that humans were protected from smallpox, that is, by vaccination with cowpox lymph.

The royal commission hired seven investigators with the stated aims of identifying the factors that promoted the spread of the cattle plague and finding a treatment or vaccine against it. Their third report of 1 May 1866 consisted mainly of the seven scientific reports. Epitomizing the culture of expertise, the commissioners admitted that this report, because of "its nature," had been prepared mainly by "the medical members of the Commission," who were relied on for their knowledge of "medicine, chemistry and physiology."[28] Each researcher had been assigned a separate project and worked independently of the others. The seven investigations, all of which were based on contemporary etiological theory and sanitary practices, focused on: (1) the nature of the disease; (2) the chemical pathology of the disease; (3) general pathology; (4) microscopic researches; (5) the application of disinfectants; (6) disinfection and ventilation; and (7) the treatment of the disease.

In this, their final report, the commissioners were happy to report that the epidemic was abating, a development they attributed to the "new repressive measures; viz. slaughter, stoppage of cattle traffic on railways, increased restriction of movement on common roads, and more complete measures of isolation and disinfection."[29] As the epidemic had progressed, these stringent measures had become acceptable in the face of the apparent alternative: wholesale destruction of British cattle with the associated social unrest.

Investigations for the Cattle Plague Commission

Two of the cattle plague scientific investigators, Burdon Sanderson and Lionel S. Beale (1828–1906), and their reports played a significant role in the later discussions of germ theories in Britain and illustrate the variety of activities that fell under the rubric of pathological research. Burdon Sanderson wrote on the "Nature, Propagation, Progress, and Symptoms of the Disease," and Beale reported his "Microscopic Researches." At this time Burdon Sanderson was merely the medical officer of health for Paddington who did some pathological research for the medical office of the Privy Council (under Simon's direction). He was also acquainted with Quain, who personally asked him to take up the inquiry. Beale was a well-respected microscopist, the author of several textbooks on the subject, and an obvious choice for his project.[30]

In November 1865 Burdon Sanderson began his research on the cattle plague with five questions in mind: Which species could be affected? What conditions allowed the plague to be transmitted through the air? Could the effects of inoculation be modified or mitigated? Where did the virus exist in the animal? And, finally, what was the protective power of vaccination? "My investigation is called that of the 'natural history' of the disease," John wrote to his sister Jane, "that is, the answer to the question what the disease is, and does when left entirely to itself."[31] Outlining the essence of his research, he explained to her that "these animals are to be observed during a period of several days in their normal condition, then inoculated with Rinderpest, then again observed until the chapter of their sad existence closes." He explained, "All about the creature-patient is to be analysed, weighed, measured or minutely described as the case may be, so as to enable the human creatures surrounding to penetrate as deeply as possible into the causes & origin of its suffering."

Inoculation was not the same as vaccination. To inoculate was to inject a source of the disease poison into a healthy patient in the hope of inducing a milder form of the disease. This procedure predated vaccination. The term *vaccination*, in contrast, referred to the application to a patient (human or otherwise) of vaccine, that is, cowpox lymph to prevent smallpox; the lymph was applied by either incision, puncture (i.e., inoculation), or scratching. In the commission reports *vaccination* sometimes also described the experimental application of the human smallpox poison to healthy cows in the hope of preventing the cattle plague. Later *vaccine* (and *vaccination*) were used to refer to any disease: the author of a 1919 textbook, R. W. Allen, defined a vaccine as any bacterial suspension in an inert fluid.[32]

Burdon Sanderson used two inoculation techniques: the insertion of a half-dozen lengths of thread, soaked with virus, and the subcutaneous injection of liquid virus under the skin of the perineum with a syringe. As a source of virus, he used various fluids from sick animals, such as the serum of blood or nose and eye discharges. He also attempted to dilute or otherwise diminish the effects of the poison and then inoculated the animal with this diluted virus in an attempt to produce a weaker, protective form of the disease.

Before inoculation Burdon Sanderson examined six animals to establish normal appearances. He took their temperature rectally, measured their urine output, examined their feces, took their pulse, measured their respiratory rates, and measured the milk output of the two cows. He also noted the appearance of the cattle's gums, lips, and other visible membranes and organs and the general behavior of the animals. After inoculation he measured, observed, and noted the changes in all these categories.

In these investigations Burdon Sanderson developed a detailed history of the disease, which included such symptoms as rashes. The commissioners hoped that farmers could diagnose affected cattle as early as possible and then quarantine or destroy the animals. Thus, they emphasized Sanderson's discovery (one I suspect was of dubious practical importance, unless Victorian farmers regularly measured their cattle's temperatures) that the earliest sign of infection was a rise of temperature forty-eight hours before any visible symptoms appeared. The animal's natural temperature of 102 °F rose to 104 °F or even 105.5 °F thirty-six to forty-eight hours after inoculation—therefore, the incubation period was significantly shorter than a week, as had been previously believed. Two days later the next symptom was "a peculiar condition of, or eruption on, the lining of the membrane of the mouth" alongside "a very distinctive appearance on the mucous membrane of the vagina." The next day the animal appeared slightly ill and to an lack appetite. It was not until the following day, or seventy-two hours after the rise in temperature, that the animal appeared decidedly ill. After the fourth day the animal's "constitution is thoroughly invaded," with a failing pulse; discharge from the eyes, nose, and mouth; and skin eruptions all evident. Death usually occurred on the seventh day after the animal's temperature had risen.[33]

The cattle plague commissioners agreed with Burdon Sanderson that his most important scientific finding was that the blood of an affected animal "contains an agent which can produce the plague in another animal." The commissioners wrote enthusiastically that this was the "most important pathological discovery yet made in Cattle Plague." They explained, "It is pregnant with consequences in medical doctrine, for though the existence of a similar fact has been long suspected in several human diseases, it has never been proved in any." This statement was slightly misleading. It had long been known that some diseases such as syphilis and smallpox could be transmitted by inoculation of the matter from sores or pustules. The distinction must be that there had never before been a successful inoculation with the blood of an infected organism.[34] Burdon Sanderson also determined that the "poison" multiplied in the blood exceedingly rapidly: in less than forty-eight hours (perhaps even far more quickly, the commissioners mused) after "a minute portion of the mucous discharge from the eyes and mouth of an animal ill with Cattle Plague" was introduced into the blood of a healthy animal, "the whole mass of blood, weighing many pounds, is infected, and every small particle of that blood contains enough poison to give the disease to another animal."[35]

In his research Beale examined microscopically "many of the different tissues and fluids of the body at different stages of the disease." In his opinion the great danger of the disease was the "highly congested state of the capillary vessels of many different textures and organs."[36] Beale stated unequivocally that "the poison itself consists of extremely minute particles of living matter, which multiply in the blood." These particles "cause local capillary obstruction, which passes on into complete stagnation."[37] The committee explained that, although it might seem easy to separate and demonstrate "the virus itself," in fact Beale had been unable to find a "definitely formed substance that can certainly be said to be the cause of the Cattle Plague."

Beale found "large collections of very simple vegetable organisms" in the center of a black clot, removed from the aorta after death. He believed that these organisms would "not live and multiply in healthy circulating blood."[38] Beale concluded that the "*contagium*" is light and can be carried by the air — though it is not volatile because it "is living matter of some kind." He felt that the "millions of contagious particles" in an unhealthy organism were "direct descendants of the very few . . . first introduced."[39] He also surmised that some who are in contact with the contagious particles escape the illness because the particles "are prevented from passing through the vascular walls, or, coming into contact with the blood, . . . they do not grow and multiply, but become destroyed."[40]

The medical press reported extensively on, and was supportive of the cattle plague investigations, in both articles and letters. The *British Medical Journal* took the opportunity to comment that the third report of the Commission investigations shows "how great are the resources of modern science, and how remarkable an increase to our knowledge of diseases generally we may expect when all have been investigated in as thorough a manner as has been the present fatal epizootic."[41] Despite such a statement, which concurred entirely with the opinions of Simon, the medical journals featured the symptoms of the cattle plague and compared them to observably similar human diseases. Thus, the *British Medical Journal* did not even mention Burdon Sanderson's discovery of "a contagious poison" and dismissed Beale's explanation for the affected animals' lung congestion.[42] In other words, when the medical journals supported pathological research into the cattle plague, they implicitly meant the collection of clinical observations. Burdon Sanderson and Beale had combined clinical observations with a variety of laboratory interventions. Their conclusions were sidestepped because much of their data was not meaningful in a medical culture not yet accustomed to the laboratory as a site of knowledge production. As will be discussed in chapter five, when the controversy over the germ theory erupted in 1875, by contrast, the debates were centered around experimental results—which had in the meantime gained more persuasive power in the medical community generally—rather than symptoms or other clinical observations.

Before the Germ Theory

The cattle plague commission's scientific reports illustrate one of the fundamental problems in understanding the era before the germ theory: terminology. At first glance words such as *germs, virus,* and *bacteria* look familiar and simple to understand. A reader might remember that viruses were not observed until the twentieth century and be a bit confused. In fact, to understand any of these words in the modern sense is to misunderstand them. During this era the meanings of words like *bacteria* and *germ* were in a state of flux (as their meanings gradually evolved into their current definitions). Not everyone's views changed at the same time, and the meaning of many pathological terms thus became uncertain. This ambiguity fueled contemporary debates about the nature of contagions. Semantic disputes were common. For example, Gerald L. Geison noted that Burdon Sanderson had argued in 1877 that Pasteur's "organized corpuscles" were not "germs" (i.e., seeds) but adult microorganisms.[43]

For a historian the first hint that the words used to describe contagious diseases were not yet clearly defined is the confusion created by attempting to map current meanings onto the past. Even more telling was the effort contemporaries made to define their terms. Their efforts and the often contradictory definitions that emerged underlined the fluidity of the field—a fluidity that was characteristic of an unformed discipline.[44]

In the mid-nineteenth century these words had multiple meanings. The first layer of meaning derived from their Greek and Latin roots, with which any educated Victorian would be familiar. A second layer of meaning was that specific to pathology. *Bacterium* was derived from the Greek word meaning "rod or staff," so at least initially bacteria were rod-shaped microorganisms, visible under a microscope. *Virus* was the Latin word meaning "poison or toxin." Pathologically, a virus was a "morbid principle or poisonous substance," a "corrosive or contagious pus" produced in the body as part of a disease process, particularly one transmitted by inoculation.[45] A germ was a sprout or a seed, therefore implicitly a living organism with the potential for further development. Thus, the anonymous writer known as the "Inquirer," in an 1877 article in the popular monthly the *Contemporary Review,* could write of "bacteria, or their germs, if germs they have."[46]

The cattle plague commissioners, for example, were well aware that pathological terms were often ambiguous. In a footnote to their third report they wrote: "The word poison, . . . is used in the general medical meaning. Unlike a chemical, corrosive, or irritant poison, it requires a second condition to be present, for it does not act unless certain favouring conditions also exist." They explained that "the terms 'germ' or 'growth' are used because no better expressions can be found. They seem to imply an independent living existence of the poison, and on this point our knowledge is not yet sufficiently definite."[47] Despite the cattle plague commissioners' clear statement that their "germs" were not the germs of the germ theory, traditionally the cattle plague scientific investigations have been seen as the turning point in the acceptance of the germ theory in Britain.

Beale and Burdon Sanderson's investigations were later constantly cited (by themselves and others) as evidence that Britain had been in the vanguard of experimental pathology. John Fisher, for example, argued that "the cattle plague forced a reevaluation of disease theory in a manner that ultimately eased the acceptance of germ theory" and asserted further that the cattle plague research brought "British science into line with Continental directions and in so doing introduced British scientists to what for many of them was a new world of achievement."[48] But, as will emerge in the discussion in the chapters that follow, the cattle plague investigations also laid the foundation for the later skepticism of Burdon Sanderson and Beale, in particular, and the British pathological community, in general, about the germ theory.

Moreover, the coupling of the cattle plague research with the experimental pathology of the 1870s and 1880s minimized the many continuities between these investigations and those that had preceded them. For example, Beale's emphasis on the cow's blood and Burdon Sanderson's use of inoculation were based on long-standing theories that diseases were due to pathologies of the body fluids and, in particular, the blood. The talk of poisons was rooted in an earlier theory—articulated by Robert Christison (1797–1882) in 1839—which connected the action of poisons to so-called morbid poisons that produced febrile diseases. Margaret Pelling has emphasized the continuity in sanitary practice between the years before 1865 and the years afterward. Nonetheless, retrospectively, and here I am emphasizing the emerging discontinuity in medical theory, it is evident that these investigations had occurred at a pivotal point in the history of etiology. By the mid-1870s Pasteur's germ theory was the focus of discussion in British scientific circles, in particular in medical circles because of the efforts by Lister to popularize it as the theoretical basis for his antiseptic surgery techniques.[49]

John Simon's Successful Sponsorship of Pathological Research

Over time Simon persuaded Parliament to require the recording of vital and disease statistics by local authorities, which made the domestic epidemiological inquiries redundant. At this point Simon attempted to persuade the Treasury that the relevant portion of his mandate could and should be interpreted to support what we would call basic research into disease causation.[50] He argued that the government had supported research into the cattle plague in 1865, and at the very least they could do as much for human maladies.[51] (During the cattle plague epidemic he had argued that the funds should be allocated for this one-time emergency.) Simon succeeded in getting funding out of the Treasury.[52]

The amount of money Simon acquired for "auxiliary scientific investigations" was not trivial. The budget was two thousand pounds, which was broken down in 1874-75 into three hundred pounds for Professor Sanderson, five hundred pounds for Dr. J. L. W. Thudichum, and twelve hundred pounds for laboratory expenses.[53] During this same period the five hundred pounds Burdon Sanderson received from the Royal Agricultural Society covered his expenses for research on large animals that required stabling and appreciable amounts of food. Burdon Sanderson's notes on anthrax experiments refer to the use of two bullocks, five heifers, two dogs, nine guinea pigs, and two cows.[54] The entire budget for Simon's department in this period was seven thousand pounds. These were relatively large sums; as a comparison, in 1876 the grant to the Royal Geographical Society was three thousand pounds.[55]

The Treasury remained dubious about these expenditures, and, although assent was given, demonstrating that the state did in fact support basic research in this period of laissez-faire, its approbation was given "hesitantly and unwillingly."[56] A memorandum in 1874 questioned whether "the results of these investigations warrant the continuance of so large an appropriation of public money." Simon strongly argued that the money was being well spent because the investigations "are of great and increasing interest to the practical objects of Medicine, preventive and curative," and the investigations are "not likely in this country to be prosecuted to any adequate extent by private medical investigators." Simon's arguments were convincing enough that, even as he was forced into retirement, the Privy Council and Treasury Department agreed that the scientific investigations were important and deserved government support. And the investigations did continue after Simon's departure. Simon's machinations came to an end in 1876 because the Local Government Board resented his high-handedness. John Lambert, secretary of the board, cheerfully pointed out to Treasury that with Simon's departure and the abolition of his position there would be "new immediate saving of £467. The ultimate saving however will be £1,800."[57]

The research done initially was organized by Simon into two main areas: the chemistry of morbid processes and the etiology of contagion. By the end of Simon's career at the Privy Council a series of investigations had been carried out on inflammation, fever, the anatomy of enteric fever and scarlatina, the etiology of cancer and the infectiveness of tumors, physiological chemistry, and the efficacy of disinfectants. These investigations had been carried out by five investigators: Burdon Sanderson, E. E. Klein (who was a histologist and later a bacteriologist), Thudichum (who continued to do research in physiological chemistry), Creighton, and Baxter. Besides these five senior investigators the budget included money for assistants, many of whom remain anonymous. Burdon Sanderson employed several men who became prominent researchers (Thomas Lauder Brunton, David Ferrier, and Klein), and Thudichum employed at least five assistants over the period, one of whom, F. J. M. Page, went on to assist Burdon Sanderson in his physiological research. Thus, through the Privy Council, Simon managed to create a community of paid pathological researchers in London.[58] Simon's influence was pivotal to Burdon Sanderson (and other aspiring researchers). Although Burdon Sanderson wanted to do research, it was the funding by the British government which permitted him to continue devoting significant amounts of time to research projects in the 1860s. The inheritance Burdon Sanderson received upon his father's death helped (see chap. 4), but Simon's patronage was also invaluable. Through Simon, Burdon Sanderson received a large portion of his income and substantial grants for research expenses. His diaries for the period included careful records of the number of hours spent working for the Privy Council, because he was paid a per diem.

Simon was an important patron outside the Privy Council as well, never missing any opportunity to further pathological research. He was involved in the decision by the University of London to dedicate the Brown Institution to pathological research, and he suggested to the Royal Commission on the Cattle Plague that the epidemic was an ideal opportunity to conduct research. Simon contributed to a new climate in London in which researchers could hope for support for pathological projects. For example, E. A. Schäfer, who had replaced Burdon Sanderson as the practical laboratory instructor at University College, asked him in 1875, "What do you think of the pathology of syphilis as a subject for a P.C. [Privy Council] investigation?"[59]

Burdon Sanderson had a vision of his own vocation and always turned to pathological and physiological investigations in his free time. Over time his relationship with Simon changed, too, and they became colleagues rather than patron and protégé. By 1874 Simon was seeking his advice about Thudichum's chemical investigations. Burdon Sanderson's assessment that Thudichum's assertions were "without sufficient foundation and therefore ought not to be made" resulted in diminished funding for Thudichum.[60]

Burdon Sanderson also worked to become a part of the London medical community. Nonetheless, Simon's influence was extremely important, not the least because he organized the research projects into small, manageable pieces that he then farmed out to the various researchers. His management contributed greatly to the "success" of the program—a success that was recognized by its continuation even after Simon's forced resignation from the Privy Council. By 1870 Burdon Sanderson had acquired enough paying research jobs to support himself. But without the patronage of John Simon and the British government at an early stage in his career Burdon Sanderson most likely would have become yet another interested clinician who had stopped doing serious research.

CHAPTER FOUR

From Clinician-Researcher to Professional Physiologist

Making the Pulse Visible

15 January 1861
 9–10 Write letters
 10–12½ Vestry [of Paddington]
 12½–1 Nil
 1–2 Writing at home
 2–4 Laboratory
 4–6 Work at Yearbook [of Medicine, Surgery, and Allied
      Sciences]
 6½ Drive
 7–8 Prepared Title of the Sheet
 8–11 Juvenile Party

26 April 1866
 Wrote Mr. Michael Frank 9–10
 Mx expt. sphygm. 10–11:30 [at Middlesex Hospital
      sphygmographic experiments]
 Brompton Hospital 12–5
 Turkish Bath 5–6:30
 RH to dinner 7–8 [Robert Haldane, his sister
      Mary's husband]
 Dr. Hudsons 8:15–12

                                   John Burdon Sanderson's diary

Burdon Sanderson kept up an exhausting pace. He usually worked twelve
to sixteen hours a day, six days a week (he tried to keep Sundays free).
He was also engaged in a wide range of activities. He had a small private medical practice, and, as a member of staff at the Middlesex and the Brompton

Hospitals, he spent many hours seeing outpatients and patients in the wards as well as performing autopsies. He was still the medical officer of health for Paddington, and he also lectured on medical jurisprudence and public health at St. Mary's Hospital. Two or three times a week he might attend a musical soirée or go on a family outing, usually in the evenings, but he often eliminated all amusements as deadlines approached. From 1860 to 1863 he was preoccupied, particularly early in the year, with his duties as an editor of the *Year-Book of Medicine, Surgery, and the Allied Sciences,* for which he prepared the sections dealing with forensic medicine and public health.[1]

In the mid-1860s Burdon Sanderson began clinical and physiological investigations into the interaction between the pulse, respiration, and the circulation. Through his friendship with Professor William Sharpey (1809–80) Burdon Sanderson was employed from 1864 onward as an assistant in the physiology laboratory at University College London, and he did some of his own research there as well. In 1866 he was appointed lecturer in physiology at the Middlesex Hospital—his first academic post as a physiologist. He was elected to the Royal Society in 1867, and in the same year, as a result of his growing reputation as a physiologist, he was invited to give its Croonian Lecture; he chose to speak about respiration and circulation.[2] The events of this chapter took place during a transitional period in Burdon Sanderson's life. In 1870 he achieved his long-term goal of moving from a medical career into full-time academic research, both physiological and pathological. Simultaneously, Burdon Sanderson developed a career as a research pathologist. From 1860 to 1864 he spent a great deal of time on the vaccination inquiry for the medical department of the Privy Council, and in 1865 he went to Dantzig to investigate an epidemic for the Privy Council and was an investigator for the Royal Commission on the Cattle Plague of 1865–66.

In snatches of spare time Burdon Sanderson continued to do his own research. He used various locations, including his home, and often his own funds for equipment and to hire assistants. He explained in 1866: "I have been converting the little room below my study into a laboratory. Vogt [an assistant] is to work there. I find it very inconvenient to have him & his work going on at Paddington, for I lose so much time in locomotion." He also used some research facilities at the Middlesex and St. Mary's Hospitals and later a laboratory he had built in rooms in Howland Street.[3]

In 1866, on an unexceptional day, Ghetal wrote to Jane Burdon Sanderson, "We have just been doing the quarterly Sanitary Report & now John is going to write about Cholera in reply to somebody's 'most preposterous notions' on the subject." She went on: "He is doing an experiment this afternoon which is somehow connected with it. Besides he is sphygmographing no end & is going to publish a 'case' in the Lancet & write the Article 'Pulse' in Dr. Reynold's book."[4]

The Medical Sciences and the Practice of Medicine

Drawing on his large and varied experience, in many public lectures during his career Burdon Sanderson explicitly outlined his beliefs and hopes for the future of a medicine based on science. Although by 1850 physiology had emerged in Continental Europe as an experimental science, with vivisection as the defining activity, Burdon Sanderson's own activities, which crossed boundaries between clinical medicine, physiology, and pathology demonstrated the fluidity of these fields. In Britain at least it was still necessary for Burdon Sanderson to justify experimentalism in medicine and to demarcate different but related territories for physiology, pathology, and clinical medicine.[5]

It was a delicate balancing act. Burdon Sanderson was not alone in constantly crossing the boundaries of these emerging disciplines while seeking to distinguish one from the other. As the events of the 1880s would make evident, others, like Huxley, had different ideas and sought to separate out physiology from medicine as a pure, biological science. The author Eliot was connected to the Huxley circle. She embedded their views in the narrative of *Middlemarch* when she portrayed Lydgate, who had published on gout, as a failed researcher. Lydgate had not succeeded in making a real discovery, meaning, implicitly, the discovery of a natural phenomena without clinical value.

Burdon Sanderson's vision was articulated in his opening address at the Middlesex Hospital Medical School in 1868.[6] He explained to the incoming class that medical knowledge was of two kinds: medical science and medical experience. Experience, which was also known as clinical medicine and surgery, was most applicable in practice. Experience was gained at the bedside, but it was not merely the observations of one practitioner. It included the mass of observations "possessed by all the medical men that now exist or that have existed in former ages." And it also included the results of postmortem examinations to confirm or correct the views that had been held during the life of the deceased with respect to symptoms and treatment. Burdon Sanderson then turned to medical science, to which he devoted the rest and indeed the large majority of his address—its relevance was, as he acknowledged, not transparent to the beginning medical student.

Medical science related exclusively to the cause and nature of disease and the modus operandi of remedies. It thus encompassed all the subjects of medical education, which, despite their complexity, formed one system of knowledge in which facts obtained from every possible source: the chemical laboratory, the ward, and the postmortem theater were studied to determine the nature and cause of disease. The science of medicine consisted of four subjects: (1) physics and chemistry; (2) descriptive anatomy, which was not a science; (3) vital physics and vital chemistry, or physiology; and (4) pathological physics and chemistry, or pathology. Burdon Sanderson had personal experience with all the medical subjects. This definition of *medical science* and the hierarchical relationship he outlined between the four subjects did not change substantially for the rest of his life. He reworked his terms and slightly altered his definitions, but in numerous addresses from this point on Burdon Sanderson was fairly consistent about the way he defined medical science, the hope it held out for medical practice, and why it should be the basis of a good medical education.

Thus, in 1880, in an address to the incoming class of University College medical students, he stated: "Everyone is familiar with what is meant by the three R's of elementary education. In like manner there are three P's in medical learning—Physiology, Pathology and Practice, which comprehend all that you can desire to carry away with you as permanent acquisitions when you leave this place."[7] Again, the relevance of practice was self-evident. He needed to explain and justify the relevance of the sciences to medicine expansively.

Three years later at Oxford, giving his inaugural address as physiology professor, Burdon Sanderson argued that a medical practitioner, although often guided "by certain technical rules . . . the embodiment of common experience" must be able to "fall back on knowledge . . . of the nature and tendency of the disease, or the action of remedies, in other words on his knowledge of pathology." There were three facets to medical practice: diagnosis, "to find out what is the matter with a sick man and foretell the issue of his disease"; therapy "to do the best that can be done to counteract it"; and prevention, "to guard against the causes of disease." For all of these he stated, "The physician who takes the highest view of his responsibilities requires to be a pathologist."[8]

But what was pathology? Pathology was "the application, not of the principles of physiology, but of its methods to the study of the causes and nature of disease and of the mode of action of remedies." Therefore, the physician who properly sought to "act with a knowledge of the consequences of his action . . . must be a physiologist."

And what was physiology? Physiology aimed to discover the relationship between the organism and its external environment. Physiology borrowed from morphology "the structure of organs" to help set problems about these organs' function—"but for the means of solving them we have recourse to the chemist and the physicist, using the means we borrow from them, just as they give them to us." To create tracings of the pulse and to explain their relationship to the circulatory system, Burdon Sanderson utilized the instruments of physics and physiology and moved among the clinic, the postmortem theater, and the laboratory. His attempt to capture graphically the pulse exemplified his philosophy of medical science.

Instruments were Burdon Sanderson's passion. He constantly acquired, modified, and used them in his varied projects. Such an inveterate tinkerer must have enjoyed the process for its own sake; in that sense his retiring temperament contributed to the development of his experimental expertise. Burdon Sanderson's personal affinity for these nonhuman, ostensibly objective observers also belonged to the broader Victorian search (rooted in utilitarian philosophy) for impartiality, consistency, and universality.

"Sphygmographing" without End

On the morning of 24 December 1864 Burdon Sanderson visited his friend and colleague "Dr. Anstie"—Francis Edmund Anstie (1833–74)—whose promising research and clinical career was cut short by an early death. That day they occupied themselves with a new clinical instrument, Marey's sphygmograph, which produced curves of the pulse. The sphygmograph was a clinical tool that could be traced back directly to the physiology laboratory.

The first mention of the sphygmograph in Burdon Sanderson's diaries are on this date, although, according to the *Lancet,* he had used the instrument since the autumn.[9]

In 1846 the young Carl Ludwig (1816–95), later head of the Leipzig Physiology Laboratory, developed the kymograph to record graphically arterial pressure in experimental animals. The pressure was measured with a metal cannula inserted into the artery of the animal; the changing pressure was conveyed via a flexible tube to a float on a mercury manometer, which would rise and fall with any change in pressure. To this float was attached a stylus, which rested on a drum covered with smoked paper. The drum was rotated uniformly by a clock mechanism. This ingenious instrument had limited clinical use, since both human patients and their researchers preferred them to survive experiments; human patients were also understandably unwilling to have a cannula inserted into their arteries. Furthermore, in this era clinicians were more interested in the pulse, a clinical entity with which they had had two millennia of experience, than in blood pressure (of which many were likely ignorant), and the kymograph apparatus did not measure the pulse accurately (although a pulse could reflect a blood pressure). Karl Vierordt, professor of physiology at Tübingen, designed the first sphygmograph. He attached the stylus to the long arm of a lever; the short arm rested on the radial artery, and thus with each beat of the pulse the stylus would mark the paper. Unfortunately, the instrument was rather unwieldy, and, by indicating mainly the frequency of the pulse and little else, it produced data poorer than any minimally experienced clinician could detect with the fingers.[10]

In his turn the Parisian physiologist Etienne-Jules Marey (1830–1904) redesigned this instrument and produced the first clinically useful version in February 1859. His sphygmograph was much smaller and lighter; it consisted of a spring with a button to be attached over the radial artery: the entire instrument could be strapped onto the arm, and it produced a continuous tracing of the changes of the pulse. The motions of the spring were conveyed to a lever with an attached stylus, which traced the movements on a strip of smoked glass that was moved by a clock mechanism located under the pen.[11]

We do not know exactly how Marey's instrument came to Burdon Sanderson's attention. The *Lancet* had reported on it in 1860, and Marey had published a large monograph in 1863 on the physiology of the circulation, which included many kymograph tracings, although Burdon Sanderson first noted reading Marey in his diary only in March 1865. Before using the sphygmograph in late 1864, he had experimented with various instruments, including the kymograph, an ear apparatus, and the laryngoscope. Already being an adherent to Marey's graphical method, upon encountering the sphygmograph, he took it up. The pulse was an important clinical indicator, but a practitioner required extensive experience to learn to read its significance with his fingertips. Marey's sphygmograph gave continuous information about the pulse, thus giving it shape. Burdon Sanderson joined a coterie of British and international researchers who hoped that these tracings would eventually produce a record of the pulse which even a neophyte could read at a glance and apply to diagnostics.

Initially, Burdon Sanderson was fascinated by the sphygmograph. In late 1864 and early 1865 he spent hours "sphygmographing." Often he experimented late into the night, sometimes alone, sometimes with his colleague Anstie; on one occasion he demonstrated it to Sharpey at University College. For Burdon Sanderson the portability of the Marey sphygmograph was vital. He worked at home, at University College, at Anstie's, and at various hospitals, where he sphygmographed his own patients and those of some of his colleagues to see the effects of pathological conditions on their pulse tracings. Burdon Sanderson stopped his sphygmographic experiments temporarily when the Cattle Plague commission research took up all his spare time from late 1865 until the middle of 1866. When that project was completed, he again sphygmographed regularly, publishing in 1867 his *Handbook of the Sphygmograph*. This monograph included the *Handbook* and a reprint of his lecture to the Royal College of Physicians, "The Mode and Duration of the Contraction of the Heart in Their Relation to the Characters of the Arterial Pulse in Health and Disease." In this booklet Burdon Sanderson promoted the instrument, even as he revealed some of its problems.[12]

Burdon Sanderson began with a Marey sphygmograph constructed by the Paris instrument maker Bregeut, which he had obtained probably through the French maker's British agent, Milne. Through experimentation he improved it. He changed the method for fixing it to the wrist, eliminating the need for the patient to remain completely immobile for good readings. He calibrated the spring to enhance the measurement of the variable "compressibility" of the artery. He substituted smoked glass for the paper as a recording device. This latter innovation had several advantages: first, since the smoked glass presented a surface with little friction, there were fewer failed attempts when the pen or ink failed to record; and, second, the tracings could be easily displayed by magic lantern or reproduced by photography.[13]

According to Burdon Sanderson, the felt pulse had four main parameters: frequency, duration, size, and compressibility. A pulse could be frequent or infrequent, depending on the number of beats per minute. This parameter could be described as well by the fingers as by the sphygmograph. A pulse was slow or quick, depending on the amount of time each beat occupied. The amount the artery dilated in length and width could be large or small. And a pulse was hard or soft depending on the compressibility of the artery. The qualities of duration and size were difficult "to appreciate" and more likely to be misdiagnosed by the traditional finger method. By contrast, they were clearly visible with instrumental aid. To detect compressibility could be learned by practitioners. Different pulses were associated with different diseases: for example, phthisis and exhaustive diseases were associated with a small pulse. A hard pulse had been an indicator for bleeding, but Burdon Sanderson had "no opportunity of investigating" a hard pulse experimentally, for by his account "nowadays it is never met with."[14]

With his work on the pulse Burdon Sanderson joined the ranks of those many Britons interested in the pulse and coronary disease—a tradition that had been strongly influenced by French clinicians and researchers, such as R. T. H. Laennec (1781–1826) and Jean Nicolas Corvisart (1755–1821). Burdon Sanderson's association with Richard Quain (1816–98), who had written the classic 1850 paper "On Fatty Diseases of the Heart," as well as his training in Edinburgh and Paris, made it likely that he was acquainted with this literature.[15] Indeed, he acknowledged such influences, referring, for example, to the classic work of William Harvey (1587–1657) and Albrecht von Haller (1708–77) and noting that, despite more recent work, such as that of J. B. Auguste Chauveau (1827–1917), "there is a constant tendency to return to the doctrines of Haller and Laennec."[16] Burdon Sanderson emphasized the British contributions to the understanding of the motions of the heart. It was Harvey "who showed that the contraction of the heart is but *one movement,* in which both auricles and ventricles take part, and that during the interval between each contraction and its successor, the heart is absolutely at rest." This observation was, he wrote, forgotten until the discovery of auscultation. Two British observers—C. J. B. Williams (1805–89), who recognized the verity of Harvey's contention and correlated the observations of auscultation, and John Reid, who gave its "most complete vindication"—played crucial roles in establishing Harvey's views.[17]

Although Burdon Sanderson did not specifically note the contributions of other compatriots to the study of the heart and circulation, Britons had been customarily more interested in the pulse than their Continental compatriots. The Dublin clinician Dominic John Corrigan (1802–80) had even had the honor of seeing the phenomena of the visible pulse due to "aortic regurgitation," dubbed the "Corrigan pulse" after his description of the phenomenon in his 1832 paper. But, as medical historian P. R. Fleming noted in his 1997 *Short History of Cardiology,* "about the pulse itself he [Corrigan] had little to say."[18]

Burdon Sanderson's work reflected contemporary concerns. His sphygmograph research could be viewed as a reply to Corvisart, who had commented that "there is something in the state of the pulse unintelligible, and which is much better felt than described."[19] Burdon Sanderson aimed to make the "unintelligible" both visible and transparent in meaning. For practitioners, as Burdon Sanderson noted, the pulse itself was not intrinsically interesting; the pulse "must be constantly studied in relation to the contracting heart which produces it."[20] As it remains today, the pulse was a means to discern the condition of the patient's circulation. In this instance Burdon Sanderson aimed to explain "the meaning of the record inscribed by the lever, or, in other words, the relation between the *form* of the tracing and the *movements* of the heart of which it is a representation."[21] He then outlined how the movement of the heart was represented by the sphygmograph tracings and how these tracings could indicate various pathological conditions. Indeed, the circulation, and thus the tracings, was also altered by the "consciousness of painful or even disagreeable impressions" and emotion.[22] The effects of the emotions on the heart had been emphasized by Corvisart, who had written, "No moral affectation can be experienced without acceleration, diminution, or other derangement of the motion of the heart." Corvisart was later criticized for emphasizing the role of the emotions, rather than purely physical causes. Despite these criticisms, researchers continued to recognize the role emotions played on the function of the heart. For example, the Glaswegian Allan Burns (1781–1813), who wrote the important book in 1809 *Observations on Some of the Most Frequent and Important Diseases of the Heart,* could not agree that mechanical causes alone could explain the dilation of the heart.[23] As Burdon Sanderson's statement demonstrated, a belief in the influence of psychic phenomena on the heart and the pulse persisted alongside more purely physiological explanations.

The Royal College lecture began with a detailed explanation of the physical dynamics of liquid expansion—a Cartesian view of the circulation, the theoretical justification of how the motion of the heart and the subsequent propulsion of blood were illustrated by the tracings of the movement of the radial artery which the sphygmograph produced. In other words Burdon Sanderson began with chemical and physical underpinnings of the movements. The sphygmograph distinguished, in his account, four events of the arterial pulsation: the primary expansion of the artery, the systolic distension of the artery, the diastolic collapse of the artery, and the diastolic expansion—when the artery "again expands after the collapse at the end of the systole." When the heart contracted, blood was projected forward into the aorta; in a similar vibratory movement, when the ventricle stopped its contraction, blood moved toward the heart. This movement of the heart expressed itself at the radial artery as a sudden expansion and a sudden collapse—indicated by the sphygmograph with a vertical ascension and descent of the recording lever. These two motions illustrated the length of systole, the time between the contraction of the ventricle and the closing of the aortic valve. In addition the tracings illustrated that these two main motions were followed by a succession of smaller movements "in alternately opposite directions." The tracings also produced evidence of absence of elasticity, which was "inseparably associated with hypertrophy and dilation of the arteries, and increased arterial resistance."[24]

Burdon Sanderson outlined how the pulse tracings should be used clinically: first, for reliability the method used must be invariable; and, second, only two important factors should be analyzed, duration and resistance to pressure. The practitioner should establish the relationship between the length of systole and diastole; the time that elapsed during the systolic distension and the gap between it and the diastolic distension; the "intensity" of the two vibrations; and "the effects of varying the tension of the spring of the sphygmograph"—which established resistance to pressure. He then turned to specific tracings to illustrate these general points.[25]

Noting that "in almost all fatal disease, the signs which indicate the ten-

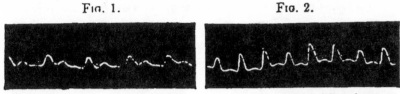

Figure 4.1. Sphygmographic tracings, Queen's Medical Photography.

dency to a fatal termination are those which imply sinking of arterial pressure and increased distension of the veins," Burdon Sanderson chose for his first two examples dying patients who were suffering from "exhaustion" (see fig. 4.1). He performed a physical examination on the two patients, obtained sphygmograph tracings, and correlated his findings with an autopsy. He stated that the tracings demonstrated that the "exhausted" heart cannot maintain the normal pressure difference between arterial and venous pressures. The first patient had had a large, harsh systolic murmur, and the hearts of both patients were obviously increased in volume. After death he found that their left ventricles were hypertrophied, unnaturally increased in mass, and the right cavities were dilated; in addition, he found venous congestion in the abdomen, which would be expected in cases of "so-called obstructive disease of the heart." He did not directly mention the state of the valves, although he noted in passing that the first patient's "loud and harsh systolic murmur" had been "erroneously supposed to indicate disease of the mitral valve." These two cases represented the weak or failing pulse. From our perspective these patients likely died of heart failure, but, even though the clinical signs of heart failure had been known since at least the time of Corvisart, the term itself did not come into use until the early twentieth century.[26]

Nonetheless, even these two sphygmograph tracings, of relatively easily diagnosed clinical problems, demonstrated that interpreting the sphygmograph records was difficult. Burdon Sanderson himself explained that, although his instructions appeared simple enough, they were difficult to carry out in practice. Echoing the language of utilitarian philosophy, Burdon Sanderson hoped pragmatically that the sphygmograph would be "an impartial and consistent witness."[27] He believed it would eventually allow young practitioners, as well as their older, more experienced colleagues, to read a pulse. But, after many hours of experience with the instrument, Burdon Sanderson was unable to produce consistent tracings; thus, he emphasized the importance of not varying the conditions under which the tracings were taken. He cautioned, "Let us, therefore, avoid being in too great a hurry to introduce the sphygmograph into the consulting-room; for if, with so imperfect a knowledge as we at present possess of first principles, we endeavour those principles in diagnosis, we shall not only discredit ourselves, but the method by which we profess to be guided."[28]

At the end of the lecture Burdon Sanderson remained guardedly optimistic. He did not believe that the sphygmograph could diagnose all circulatory conditions better than a practitioner could without its assistance. He predicted that the sphygmograph would be of invaluable diagnostic use in providing information about "the probable duration of life." He explained that for many apparently healthy people the tracings indicated excessive arterial resistance. He concluded: "The questions is, Are such persons sound? I think not; but, as yet, we know too little to speak dogmatically."

As Jacalyn Duffin has stated, the early-nineteenth-century interest in the pathology of the heart could be summarized by four points: changes in the heart caused illness; these organic changes had origins that were both physiological and psychic; studies of the physical lesions emphasized the enlargement of the heart and its muscle; and clinical signs were few and frequently depended on the patient's perception and description.[29] Burdon Sanderson's research and writings, although lacking any acknowledgment of many of his nineteenth-century predecessors, had many similarities to their work—in particular, his focus on the heart and his continued belief in the impact of emotions on the heart and circulation. His comments about high "arterial resistance"—perhaps indicating an inkling of what we would call hypertension—symbolized a change that occurred in the first half of the nineteenth century; cardiac researchers turned their attention to the valves of the heart and the arteries. Nonetheless, the continuities are most striking. The diagnostic revolution of the early nineteenth century, symbolized by the stethoscope, had changed practice, but clinicians, perhaps particularly British ones, still hankered after clearer, more reliable clinical signs of cardiac illness.

Experiments on Respiration and the Circulation

In 1867 Burdon Sanderson was pleased and honored to be elected to the Royal Society and asked to give its Croonian lecture. Once the date for the lecture was set, he cleared his schedule as much as he could and worked up a whole new series of experiments. Although the experiments were new, this lecture, "The Influence Exerted by the Movements of Respiration on the Circulation of the Blood," was the cumulation of many years of work on the subject. In a related set of experiments Burdon Sanderson had earlier tested two techniques for the revival of drowning victims, using mainly dogs as animal models.[30] The published paper in which Burdon Sanderson, for a medical audience, nonchalantly referred to the struggles, gasping, and suffocation of an unnamed number of dogs demonstrated an attitude that would prove to be out of step with British public opinion.

There is a direct relationship between these animal experiments—his physiological research—and his clinical work on the pulse. He carried out these physiological experiments on dogs during the same period as the sphygmograph experiments on human objects, and in his clinical papers on the pulse he referred to this Croonian lecture. The connection was equally evident to his Royal Society audience. Burdon Sanderson began the lecture with a quotation from Carpenter's *Physiology:* "During the act of expiration, the frequency of the pulse is considerably augmented, whilst the line of mean pressure rapidly rises, indicating increased tension in the arterial walls." "During the act of inspiration," in contrast the text stated, "the pulsation becomes slower, the curves much bolder, and the line of mean pressure gradually falls; for then the blood readily enters the thorax, and as a consequence, the great veins, capillaries, and arterial walls become relatively flaccid." Burdon Sanderson disagreed with these conclusions. Through his experiments on dogs he contended that in normal breathing inspiration resulted in an increase in the force and frequency of the contractions of the heart: "the commencement of inspiration being immediately followed by an increase of pressure, which becomes still more marked during expiration, but again subsides at its completion." This phenomena remains interesting. In the classic *Guide to Physical Examination and History Taking* clinician Barbara

Bates described how blood pressure levels fluctuate during any twenty-four-hour period, due to activity levels; emotional state; pain; temperature; coffee, tobacco, and other drugs; and even simply the time of day.[31]

Since they were animal experiments, Burdon Sanderson was not restricted to the use of a sphygmograph. He used two instruments: one to measure the animal's respiration and another to measure the arterial pressure. The dog was forced to breath through a T-shaped tube, by attaching the tube either to the trachea or to the mouth with a mask over the snout. One arm of the tube was left open to the air; the second was connected to a "disc-shaped bag of thin caoutchouc." This bag expanded and collapsed as the animal breathed; to its upper surface was attached to a vertical rod. The arterial pressure was indicated by a mercurial manometer, inserted into the femoral artery, and on the surface of the mercury was a cork float, which was also attached to a vertical rod. The movements of these rods—one monitoring inspiration and expiration, the other changes in arterial pressure—were then recorded, using a system of levers attached to fine sable brushes on a roll of paper which was moved by clockwork.

Burdon Sanderson began each experiment by recording "normal respiration." He also varied the width of the T-tube opening in order to see whether the resistance of the T-tube to the passage of air affected his results. After this series of experiments, using the same animal "under the influence of morphia," he concluded that this resistance did not affect his results. He then plugged the trachea and used a mercurial manometer in place of the caoutchouc bag to demonstrate that, when respiration was blocked, the relation between the respiratory movements of the chest and the arterial pressure was the reverse of that described earlier. Using other dogs, Burdon Sanderson explored the influence of chemical, mechanical, and nervous factors in ordinary breathing. He concluded that "the influence of the thoracic movements on those of the heart may be either directly mechanical, as in suffocation, indirectly mechanical, as in ordinary breathing, or chemical."[32]

These experiments were simple in conception but difficult to execute. They also depended, as Burdon Sanderson explained them, on an understanding of the chemical processes—he integrated references to the "reaction of oxygen on the circulating blood"—and physical processes underpinning respiration.[33] There was a synergy between these experiments and the sphygmograph experiments. Both contributed to Burdon Sanderson's evolving understanding of the motions of the heart, the circulation of the blood, and respiration.

Physiology and Pathology: New Career Directions

Burdon Sanderson continued sphygmographing intermittently during 1868 and 1869, even though much of his time was occupied with doing pathological researches for the Privy Council. Around this time there was much interest in the instrument; for example, there were occasional questions in the examinations for members of the Royal College of Physicians on the form of the pulse in health and disease, as demonstrated by the sphygmograph,[34] and in 1868 George Johnson of King's College cited Burdon Sanderson's sphygmographic evidence of "increased arterial pressure" in an article on Bright's disease.[35] Burdon Sanderson's patronage of the sphygmograph firmly established it in Britain, as Robert Frank described, and pulse tracings were used to illustrate particular pulses in publications for the rest of the century. In addition, the later work of James Mackenzie (1853–1925), who in 1902 published *The Study of the Pulse*—in his emphasis on function over structure and use of "a polygraph of his own invention"—demonstrated some similarities to Burdon Sanderson's work. Mackenzie was strongly influenced by the physiologist W. H. Gaskell (1847–1914), who was a friend and colleague of Burdon Sanderson.[36]

As for Burdon Sanderson he referred to the instrument in his section of the *Handbook for the Physiological Laboratory*, published in 1873, but after the 1867 publication of his guide to the instrument his initial euphoria diminished. This disillusionment was not unique. Although by the end of the century the transfer of instruments and ideas from the physiological laboratory (evident in the sphygmograph, the sphgymomanometer, the electrocardiograph, and X rays) created new categories of heart disease, which fueled the "new cardiology" of the early twentieth century,[37] the sphygmograph never really succeeded as a clinical instrument in practice.[38] Perhaps that is why Ghetal did not even mention this episode in her *Memoir*. She was well aware of this research, having assisted in sphygmographic experiments and having been intimately involved in the production of the book. She had helped John with the necessary varnishing of the tracings and had accompanied him on his visits to the lithographer and printer.

Nonetheless, the sphygmograph and more successful instruments of the mid-nineteenth century, such as the thermometer, demonstrated that physiology had the possibility of materially and substantially altering medical practice for the better. In addition we can see why Burdon Sanderson believed that successful diagnostics required an understanding of physics and chemistry and an intimate knowledge of physiology and pathology. He began with Marey's instrument but then modified it in important ways to improve its reliability; these modifications demonstrated his grasp of basic physics as well as the technical requirements of instrument making. To make the tracings intelligible, he also utilized a physical understanding of the motion of liquids; the act of respiration itself had chemical components. The pulse tracings themselves led him first, as it had the anatomical pathological school of Paris, to attempt to correlate these exterior signs with the inside of the body through postmortem autopsies. Burdon Sanderson had already been interested in circulation and respiration (in part that was why he had taken up the drowning inquiry), but the tracings of human pulse which the sphygmograph produced reinforced this interest. He moved between his tracings and animal experiments in an attempt to understand more fully the phenomena of circulation and respiration, which he hoped the pulse tracings more concisely embodied. In the sphygmograph we can see the intersection of Burdon Sanderson's clinical and physiological interests. Unlike his Edinburgh professor, Bennet, Burdon Sanderson was able to utilize his patients, autopsies, and physiological experiments together.

In 1867, the same year that he published his handbook on the sphygmograph and delivered the Croonian lecture, Burdon Sanderson resigned as medical officer of health for Paddington. He found it increasingly difficult to carry out his duties because he was now very occupied with pathological research for the Privy Council, in addition to all his other activities. Some officers of the vestry were irritated by the amount of other work Burdon Sanderson had taken on. His medical officer of health salary was to be replaced by contract earnings from the Privy Council, a less reliable source of income, but John and Ghetal decided that they could now afford the risk, since John had inherited some money upon his father's death in 1867.

Their gamble paid off in 1870, when Burdon Sanderson was appointed professor of practical physiology and histology at University College London, replacing Michael Foster, who had moved to Cambridge. He was now secure enough to resign his clinical positions and effectively end his private practice—making an exception in 1873 for the autopsy of the "late Emperor of the French," an honor doubtless due to his renown as a pathologist. In part Burdon Sanderson gave up his career in clinical and preventive medicine because of what he perceived to be his greater talent and interest in scientific research. But these resignations also demonstrated his hierarchical view of the relationship between medical science and practice; practice should be based on science, which thus should have a special and higher status. His decision also reflected the growth of the fields of physiology and pathology in this era. It was no longer practicable to combine clinical practice with an active career in physiology or pathology. In order to be a successful experimenter, Burdon Sanderson needed to devote more and more time to the acquisition of the tools and techniques of research, which involved much travel, reading, and interaction with colleagues at home and abroad. He had never really had the time to build up a private practice, and now even attending to hospital patients was difficult. Indeed, some of his British colleagues felt that it was already impossible to maintain parallel careers in research pathology and physiology; they advised him to focus his energies fully on one or the other.

In 1871 University College London decided to fulfill the conditions of a bequest that had been made in 1852 by a Mr. Thomas Brown, one typical of the private support for research of the era. Brown had willed a sum to the university "for the founding, establishing, and upholding an Institution for investigating, studying and ... endeavouring to cure maladies, distempers, and injuries" of "Quadrupeds or Birds useful to man." With the encouragement of Simon (among others) the university decided to use the money to establish a pathological laboratory. After a long period of delay, the university suddenly sprang into action because Brown had specified that, if his wishes were not fulfilled within nineteen years, the money was to go to the University of Dublin. Burdon Sanderson was closely involved in the plans for a building, and he was appointed the first professor-superintendent of the Brown Institution. The choice of Sanderson was unsurprising given that the members of the Committee of Directors were: James Paget, John Simon, Dr. Buchanan, William Gull, Richard Quain, and Professor Sharpey.[39] He closed his private laboratory in Howland Street, where he had carried out most of his investigations for the Privy Council on the nature of contagion.

Thus, by 1870 Burdon Sanderson was effectively both a physiology and pathology professor. He was already deeply involved in the plans for the Brown Institution, although he had not been officially appointed professor-superintendent. His attainment of these positions marked a new era. Although he was a seasoned experimenter who had organized many projects, worked in different locations, and had already designed several laboratories, the responsibilities of being professor made him busier than ever before. In addition to his lectures and administrative duties, he supervised the laboratory at University College and its renovations, along with the construction of the Brown Institution. Alongside his own programs of physiological and pathological research, he met with students to oversee their research projects. With these appointments Burdon Sanderson had achieved his goal of becoming a full-time medical researcher, but in the end, ironically, being engulfed in his new administrative responsibilities and dogged by ill heath, he would have less time for his own research.

Burdon Sanderson's new ability to make a career in research was part of a wider phenomenon. He was not alone—Foster's move to Cambridge, for example, demonstrated that other colleagues were equally successful. The next generation of would-be researchers would be able to vie for a number of posts. In the 1860s (indeed, as Eliot created the portrait of Lydgate in 1866–67) it was not yet evident that future Lydgates would have more opportunities than their grandfathers had had in the 1830s. The 1870s were a turning point in Britain. Although research-minded Brits still looked to the Continent for direction, after the 1870s the number of men in Britain engaged in serious biological research had reached a critical mass.

CHAPTER FIVE

Becoming a Research Pathologist

The Rise of Laboratory Medicine in Britain

[John] says that he is convinced that the epidemic he is going
after [cerebro-spinal meningitis] is not infectious whatever any
one may say. I don't know how he already knows this, but it is at
any rate consoling to believe it.

Ghetal Burdon Sanderson, 1865

Poor Ghetal has been confined to bed for the last week — Ever
since the date of her last letter to you when she was beginning to
feel ill — with smallpox. I had been experimenting for the
Committee on the inoculation of smallpox virus to cattle &
brought it home with me from the smallpox hospital.

John Burdon Sanderson, 1866

The events of the 1870s, the experiments, the publications, the public and private debates, transformed the rather diffuse beliefs about the causes of disease of the late 1860s into the germ theory of the 1880s. Germs were not discovered, or seen for the first time in this period. Their construction did not involve their literal manufacture, but they were the result of a process that involved the exchange of materials, ideas, and techniques among researchers.[1] Burdon Sanderson participated in these exchanges and at the same time established his international reputation as a research pathologist. The recognition of germs as pathological agents also depended on gaining consensus from the medical community and even the public at large. Thus, the acceptance of the germ theory rested on the acceptance of the utility of experiments in general for etiology, and thus medical practice, and the acceptance of the results of specific experiments.

In the mid-nineteenth century the relevance of experiments to medicine was contentious. There were many researchers, especially in France and Germany, who argued that the practice of medicine should be based on the knowledge generated by experiments and, in particular, vivisection.[2] The development of the germ theory is a case study in the wider adoption of these research ideals in Britain, first among a small group of men interested in pathology, then within the broader medical and pathological community, and finally by the educated public. Thus, with the germ theory we can see the movement of the ideals of research physiology into the medical mainstream. Even as these ideals were becoming more accepted in the medical community, a group actively opposed to them emerged from the public at large — the antivivisectionists.

The locations of these experiments, as Burdon Sanderson's activities illustrated, were diverse: homes, specialized research facilities such as the Brown Institution in London, stables, hospitals, university laboratories, and so on. The important distinction was that this knowledge was gained through experiment, not through clinical experience at the bedside. The relative importance of experimental or laboratory knowledge versus clinical knowledge ebbed and flowed in this era. The emergence of germs, which through the discipline of bacteriology came to define the science of pathology in the last two decades of the nineteenth century, depended on the rise of laboratory medicine. Underlying the construction of germs was a fundamental change in the types of explanation which medical practitioners found acceptable for disease causation. Germs were accepted because the educated medical community found laboratory data more compelling than clinical evidence.

Here the emergence of germs will be analyzed by focusing on Burdon Sanderson's research for the Privy Council and the controversy within British scientific circles in the 1870s about the nature of "ferments and germs" — most famously illustrated by the germ theory debate at the London Pathological Society in April 1875, in which Burdon Sanderson was an important participant. It was only in retrospect that this controversy gained particular significance. "Germs," "bacteria," and "contagions" as such were never the focus of much medical discourse in mid-Victorian Britain. This particular germ theory controversy began slowly in the late 1860s, peaked around 1875, and then seemed to disappear almost entirely.

Although these debates did not assume the weight they are often ascribed in the present historiography, they do reveal much about the nature of Victorian medical belief and the process by which germs were eventually accepted in Britain. These debates were due in part to British nationalism, turf battles within the biological sciences, and the powerful personalities involved; in particular, the debates, alongside the intersecting controversy over spontaneous generation, revealed the splinter between medical circles and Huxley's group.

The germ theory continued to be discussed even when the British controversy seemed to have been resolved in favor of either a physico-chemical theory of disease causation or a modified germ theory that included other environmental factors. Despite its various and competing incarnations, the idea of one germ theory remained appealing. These debates illustrated the growing importance of the laboratory within the medical profession.

When speaking of the germ theory, historians have tended to treat laboratory techniques and etiological theory as unproblematic. Since there were numerous germ theories under discussion in the mid-Victorian period, the use of the expression *the germ theory* (or its "reception") is itself somewhat presentist. In the 1870s the experimental data, by all accounts, were not conclusive, and other forms of knowledge, notably clinical experience, provided equally inconclusive observations.[3] Nonetheless, even in the face of such contradictory evidence, and with conflicting theories about the relationship between microorganisms and disease processes, it is not entirely presentist to refer to the germ theory because the expression was widely used at the time.

The initial appeal of the germ theory to medical practitioners, or at least those interested in theories of etiology, could be described as almost an expression of faith; that is, the appeal was not based entirely on tangible evidence: to believe in the germ theory was to be optimistic. Ironically, even as the laboratory evidence became more compelling, practitioners became somewhat disillusioned, as revolutionary new therapies did not immediately emerge from the many new discoveries of and about microbes. But their disillusionment was muted, since the study of germs was now within the new domain of bacteriology. Bacteriology had become an established research area with its own audience.

Burdon Sanderson's Pathological Research for the Privy Council
Experiments on the Nature of Contagia

Burdon Sanderson's research, begun in 1865 for Simon, was a signifi-cant departure for the Medical Department of the Privy Council. Burdon Sanderson repeated Jean-Antoine Villemin's (1827–92) recent experiments on the communicability of the tubercle by inoculation. Although novel for the Medical Department of the Privy Council, this line of research resembled the work Burdon Sanderson had carried out for the Royal Commission on the Cattle Plague. He aimed to determine if the "tubercular disease" was zymotic, as Villemin now claimed, or whether it was a constitutional disease. As discussed in chapter 3, zymotic diseases were caused when a particular animal poison (for a given disease) entered the bloodstream of a person who was susceptible. It was clear that the suggestion for this research came from Simon. Burdon Sanderson noted that it was Simon who had communicated Villemin's results to the Pathological Society of London. Villemin's research has been traditionally cited as the first identification of a microorganism caus-ing tuberculosis prior to Koch's 1882 work.[4]

For his study Burdon Sanderson used fifty-three guinea pigs. The first thirty were inoculated "directly with diseased material from tuberculous patients," then of the remaining twenty-three "thirteen were inoculated di-rectly from animals of the first series, which in their turn yielded material for the inoculation of the remainder." After some animals from each series had died or been killed, Burdon Sanderson filed his report. He described both the gross anatomical features of the dead animals and the microscopic anatomy, using histological techniques. Then he reviewed the relevant litera-ture, beginning with the experiments of Villemin.[5]

Burdon Sanderson aimed to uncover disease process and to confirm that the animals inoculated with the tuberculosis matter would develop tubercu-losis. Initially, he was successful; in a postscript, however, he expressed doubt about the relationship between inoculation and the disease. In some cases he concluded that the injection itself had caused the tuberculosis: "In this case, therefore, a pathological process which had originated in traumatic subcu-taneous suppuration resulted in lesions which were characteristically tuberculous." This condition he dubbed traumatic tuberculosis. In a follow-up report the following year he concluded tentatively that the idea of a "specific contagium of tubercle . . . is not as yet disproved by the facts of traumatic tuberculosis." The disease described here was called both "tubercle" and "tuberculosis."[6]

In 1869, as Simon reported, Burdon Sanderson took a new tack: "We believe that at last it has become possible, with the assistance of the microscope, to make direct studies of the intimate nature and natural history of the contagia." They hoped to contribute to knowledge about zymotic diseases, a domain that Simon contended was larger than previously understood and perhaps included cancer and tubercle. Simon also associated Burdon Sanderson's results with Pasteur's and Heinrich Schröder's views "on the agencies of fermentation and putrefaction" and concluded that their views "are importantly strengthened by the evidence which M. Chauveau and Dr. Sanderson supply . . . as to the essentially particulate form in which the morbid ferments exist and multiply."[7]

Burdon Sanderson divided his studies of the "Intimate Pathology of Contagion" into three parts. The first, related to "such physical qualities as fluidity, volatility, density, solubility; the second to the presence or absence of that orderly succession of changes of form which distinguishes living from dead organic substances." The third, which he left for a later date, was "the determination of the chemical composition of contagious matter."[8]

For these investigations Burdon Sanderson repeated the work of Chauveau, who had also carried out the cardiac research discussed in chapter 4.[9] Chauveau had studied the relationship between cowpox and smallpox and simultaneously developed techniques for isolating the different segments of the vaccine lymph, which was made up of minute particles, leukocytes, and clear liquid. His techniques, which Burdon Sanderson further refined, were simple in conception. Using "the method of diffusion," Burdon Sanderson separated out the parts of the lymph by preferentially dissolving the vaccine's soluble products, carefully removing the dissolved upper layer, and inoculating (and later studying) the remainder. Because of the small amounts of lymph involved, the technique was tricky and required manual dexterity. In order to learn Chauveau's method, Burdon Sanderson traveled to Lyons and assisted Chauveau himself. Even after this personal contact, Burdon Sanderson continued to write to Chauveau for advice as he carried out his investigations.

With this technique Burdon Sanderson isolated the contagious matter and studied it physically, chemically, and microscopically. "The fundamental inference," he concluded, "is that every kind of contagium consists of particles." In addition contagious fluids contained organic forms: "ferment plants . . . which have no direct relation to our present inquiry" and microzymes, which are initially spherical but become rod shaped and are called "bacteridia" or "bacteria." These microzymes are "associated with the commencement of decomposition of nitrogenous compounds of which putrefaction is the continuation."[10]

Burdon Sanderson partly sidestepped the issue of spontaneous generation for microzymes when he declared it irrelevant to pathology. He emphasized that they usually arose by reproduction and further explained that some botanists felt that they were "a race of more or less similar individuals." Ernst Hans Hallier (1831–1904), a German researcher, in contrast, posited that the microzyme colony sprang from "the reproductive filaments of a fungus of higher organization" and could develop into that same fungus under the right conditions.[11] Burdon Sanderson disagreed. He believed that there was not yet enough evidence to decide and that more research was needed.

What was the relationship between contagium particles and the smallest known organisms? If the microzymes, as Burdon Sanderson called them, were the contagia, different species of microzymes should exist to correspond with different contagious diseases. Hallier had tried to establish the specificity of microzymes, but Burdon Sanderson was not convinced. He suggested that "all microzymes are not contagium particles, yet all contagium particles may be microzymes."[12]

In his next investigation Burdon Sanderson cultivated, as he put it, his own "microzymes (bacteria)" to explore what conditions aided or impeded their development and defined microzymes—"living particles which in their earliest state are spheroids, and do not exceed 1/20,000 of an inch in diameter, but subsequently elongate into rods." Microzymes were assumed to be albuminous (because they stained like albuminous substances) and were chemically indistinguishable from one another. They were "the universal destroyers of nitrogenous substances . . . the pioneers if not the producers of putrefaction." Burdon Sanderson concluded that Hallier was incorrect: fungi were not developed from microzymes, since the conditions in which fungi thrived were not amenable to microzymes. Burdon Sanderson was not the only researcher to question Hallier's assertions: Anton de Bary (in 1868) and Ferdinand Cohn (in 1872) also challenged his conclusions.[13]

In a pioneering study Burdon Sanderson established experimentally that water and moist substances contained microzymes. He challenged the assertion of contemporaries that bacteria floated freely in the air, contending that bacteria were only found in the air's moisture. He also tested the zymotic property of water and body tissues. To have the zymotic property was not necessarily to contain microzymes, although that was sufficient. A solution was zymotic if a small amount added to Pasteur's solution produced cloudiness after a period of days, even if "such waters often contain no elements or particles whatever which can be detected by the microscope." Burdon Sanderson tested body fluids for the zymotic property because he was convinced that no visible microzymes existed in the fluids of persons affected with contagious diseases. With his usual thoroughness he acknowledged that his failure to find microzymes did not mean there were no other germinal particles, undetectable by the microscope. He concluded that normal fluids were not zymotic and that the liquid products of inflammation were occasionally zymotic. Burdon Sanderson did not explain whether the hazy Pasteur's solution always contained microzymes.[14]

Experiments on Fever and Inflammation

From 1872 to 1875 Burdon Sanderson moved from the study of contagia into the relationship between those organisms and the pathological processes of fever and inflammation. Doubtless, Burdon Sanderson had recent experience with these pathological conditions in the clinic, as a hospital physician and as medical officer of health. In addition the study of inflammation, like that of contagia, was prevalent among researchers on the Continent during this period. His study of fever consisted of "an exposition of the clinical and experimental observations." He left "the work of combining them into a theory of fever to a future opportunity." The observations were carried out using a calorimeter and fevered dogs (the fever was induced by the injection of pus) to determine the heat production over a period of days in comparison to normal dogs. He devoted the bulk of the paper to a survey of the experimental literature on fever, concluding that the origin of fever was still unknown, although "at bottom we are all humoralists and believe in infection."[15]

Inflammation, the response of a tissue to either mechanical or chemical destruction, was a common pathological condition, the result of both accidents and surgery. Burdon Sanderson carried out extensive research on infective inflammations. Here he used the contemporary surgical distinction. A noninfective inflammation produced laudable pus, was "limited in duration and extent by the limits of the injury which caused it," and was associated with a slight febrile disturbance. An infective inflammation produced "irritative fever," "pyæmia," and "septicæmia," three states in which the inflammation "spreads and endures beyond the direct and primary operation of its cause." These three pathological states were not clearly differentiated but represented a continuum of inflammation along which irritative fever was the least dangerous and septicemia the most.[16]

Burdon Sanderson reminded readers that his previous research (on the tubercle) had demonstrated that when inflammations were induced in lower animals there were two possible outcomes, both infective: a chronic disease resembling tuberculosis; or an acute disease, pyemia, which usually terminated in death. "To a certain extent," however, the experimenter could produce either outcome at will. The acute pyemic outcome was produced by increasing the quantity of infective material inserted by ten or twenty times (the most important factor) and by generally employing material from an acute infection.[17]

Burdon Sanderson asserted "that if infective agents are particulate they are probably comprised in that group of bodies to which I then applied the term microzyme." The term *microzyme,* as he carefully explained, meant the *zooglaea* of Cohn, the micrococci of Hallier, and the forms described as bacterium and vibrio. He acknowledged two new facts about these organisms: (1) all acute infective inflammations contain microzymes in their exudation liquids; and (2) the same forms are found in the blood of the infected animals.[18]

Pyemic and septicemic diseases were caused by a "morbid poison or contagium," which he found, "so far as can yet be discerned, is a particulate ferment of ordinary putrefaction." Those diseases were produced by this contagion's intense action, the contagion produced "by less intense action the chronic tubercular infection of the body."[19]

In animals suffering from the acute infection Burdon Sanderson had found microzymes in the "exudation-liquids of the inflamed parts." Inflammations did not need to be started by a contagion from a preexisting case of the same type or by inoculation with a foreign putrid matter but might, he demonstrated, "be highly infective merely in accompaniment of being highly intense." He showed that the "compound quality of infectiveness and intensity inflammation" may be artificially cultivated; starting with a purely chemical lesion, through successive inoculations from subject to subject one can derive "one of the most tremendous morbid poisons which the mind of the pathologist can conceive."[20]

What was the relationship between the visible microzymes and contagions? Burdon Sanderson was not sure. Despite his statement that they were equivalent, he tended to believe that the microzyme carried the contagion into the body but was not itself the contagion.[21] In his second report on the same subject he wrote, "The presence of characteristic organic forms in infective liquids, affords *in itself* no conclusive evidence that these bodies are themselves the cause of the infectiveness." On the other hand, it was extremely unlikely that these organic forms had no pathological significance: "If these infinitely minute organisms are present in every intensely infective inflammation, we may be quite sure that they stand in important relation to the morbid process."[22]

Burdon Sanderson continued to experiment with "the common ferment of putrid infusions." The ferment both excited "putrefaction in dead animal matter" and had "very definite toxical relations" to the living animal body. He demonstrated that, if such an infusion were passed "under pressure through porous porcelain," the filtrate, now devoid of microphytic organisms, will not "undergo any further zymotic process" or produce in the living body any "morbific influence."[23]

Because the septic ferment could be separated by filtration through porcelain, it was particulate. But Burdon Sanderson now believed that he and Peter Ludwig Panum (1820–85), the Danish physiologist who had studied with Bernard, had proven that it did not consist of living organisms. Panum, like Burdon Sanderson, had a wide-ranging career. After studying with Bernard, Panum became professor of physiology in Kiel, Schleswig-Holstein, from 1853 to 1864; during this period he became known for his work on blood transfusions, embolisms, and blood volume. When his Danish nationality became problematic (as Schleswig-Holstein was forcibly moved from the Danish to the Prussian sphere), he moved to Copenhagen, where he organized the first physiology laboratory in Denmark. Noted for his reports on a measles epidemic in an extremely isolated community, Panum had looked also at the question of whether the microscopical organisms present in putrid liquids were necessary to produce the symptoms associated with infection. He concluded that they were not.[24]

Burdon Sanderson wrote that Panum had performed his experiments at a time "when the 'germ theory' had not been heard of, that no one had written about 'microzymes,' and that bacteria were familiar only to morphologists." Panum had utilized techniques that would be expected to destroy any living organism: for example, boiling, drying, filtration, digestion with alcohol. Even after such treatment the infusion induced illness when inoculated into an animal.[25] As a result, Burdon Sanderson concluded that infective inflammations were not caused by a live contagion, but he still believed that contagious diseases were caused by living organisms. He had earlier separated inflammations from infectious diseases.[26]

In summarizing these experiments, I have tried as much as possible to preserve the flavor of Burdon Sanderson's own changing terminology, which demonstrated the tentative quality of many conclusions in the field as a whole. For Burdon Sanderson, in particular, the words for the observed phenomena changed because Sanderson's own views were constantly evolving in the face of new experiments and new data.

A comparison of Burdon Sanderson's medical and scientific publications further demonstrated his view of the proper relationship between the experimental sciences and the practice of medicine. In his scientific publications he tended to separate the scientific/laboratory domain from the clinical, using only experimental evidence to support his assertions. Even in his scientific publications he conceded that pathology was supposed to be practical, a reminder that it should explain the phenomena doctors encountered and eventually offer solutions to their problems. For medical audiences, by contrast, Burdon Sanderson explicitly connected the two domains. In that case he would refer to clinical phenomena to support his experimental results, and vice versa. He supported his belief that pyemia could be produced by external circumstances (which he had done experimentally) by stating, "Many such cases as these are on record, in which there is not even the possibility that the disease can have arisen from without." His belief in the infective quality of pyemia was further reinforced by the well-known spread of such infections in hospitals.[27]

Burdon Sanderson's constant references to the evidence derived from experiment was characteristic of his scientific style. He intensely disliked theoretical discussions. During the 1875 germ theory debate he hoped aloud that it was possible to "put these theoretical questions into the cupboard, and turn the key upon them."[28] His dislike of theorizing reflected his belief, reiterated in most of his pathological papers, that the objects of pathology were practical—that is, related to the eradication of disease. Questions such as the taxonomy of bacteria could therefore be left to morphologists.

This work was representative of Burdon Sanderson's style of research in other ways. The design of the experiments was highly derivative; in many of them he simply repeated the research of others. He became personally acquainted with many major figures in the study of contagion, including Theodor Billroth (1829–94), Chauveau, and Max von Pettenkofer (1818–1901, a friend of Simon).[29] During this period Burdon Sanderson was in the forefront of research pathology. His pathological experiments ably displayed his prowess as an experimenter; familiar with the international literature, he had mastered the major techniques in the field.

The Controversy over the Germ Theory in the 1870s

An opponent of the germ theory, H. Charlton Bastian (1837–1915), professor of pathological anatomy, University College London, opened a debate on the germ theory of disease at the London Pathological Society on 6 April 1875. "The subject of the relation of the lower organisms to disease," Bastian asserted, "has a growing importance. The notion that there is a distinct causal relation between the two—though it has long existed in one form or another—is one which has spread enormously within the last few years, partly owing to our increase of knowledge concerning the low organisms in question, and partly because of their ascertained presence in numerous diseased tissues and exudations." After Bastian's address, Burdon Sanderson was asked to respond, and then the floor was opened for discussion.[30]

Burdon Sanderson muddied the waters when he explained that in 1870 he had contended that, in those diseases in which "minute, apparently, living particles" were found, these particles were independent of the tissues where they were found; here he opposed Beale's contention that the particles originated in the diseased tissues. Neither stance required the observer to conclude that the particles necessarily caused the disease. More important, Burdon Sanderson's opening remarks demonstrated that there was no consensus about the nature of these particles or whether they were germs of disease. In fact, despite his (and the other participants' at the meeting) constant use of the expression *the germ theory* a single such theory did not really exist in the 1870s; there were many germ theories. This plurality was illustrated by the title John Drysdale—a provincial physician who was not a member of the London Pathological Society—chose for his 1878 booklet, *The Germ Theories of Disease.*[31]

The general acceptance of the expression *the germ theory* signified the growing acceptance of the theory itself, or at least the growing hope for an acceptable theory. By setting up the terms of the debate as the germ theory or not—that is, at least implicitly, the germ theory or nothing—its proponents had already scored a coup. From our perspective it is evident that the participants saw themselves as engaged in mere controversy, but they could also be described as engaged in the process of generating a consensus about the nature of disease, in effect of creating germs, rather than simply debating the germ theory.

At the Pathological Society meeting Burdon Sanderson argued that, if you were to ask any contemporary eminent pathologist, "Do you believe in the germ theory?" the answer would be, "I really cannot give you any opinion upon the subject." He elaborated, "A great number of observations have been made upon the subject; you must read those observations; then, if you wish to pursue it, you must make observations yourselves, and perhaps at a future time it may be possible to come to a conclusion upon the subject." If pressed, Burdon Sanderson asserted, "I believe all of those eminent men would shrug their shoulders."[32]

As we have seen, Burdon Sanderson himself believed at this time that the contagions involved in infectious diseases were particulate, self-duplicating, and alive and likely to be bacteria or microzymes, as he preferred to call them.[33] He also believed that inflammation consisted of two distinct phenomena; one infective, one not. Infective inflammations were caused by contagions that were particulate but not living. The infection was carried by

either the lymphatic system or the veins. The related phenomenon of putrefaction (inflammations could become putrid) was associated with bacteria.
But it was not clear whether bacteria were the cause of putrefaction or merely
derived nitrogen and carbon from its products. Admitting that some might
find this position vague, Burdon Sanderson stated, "So long as uncertainty
exists, there is nothing to be so much avoided as that sort of clearness which
consists in concealing difficulties and overlooking ambiguities."[34]

Since he could not answer unequivocally, "No, I do not believe in the germ
theory"—and because he rejected any theory that depended on the spontaneous generation of bacteria—Burdon Sanderson was usually considered,
along with Lister, as one of the leading exponents of the germ theory in Britain. But he was also claimed by the opposition. In the Pathological Society
debate, for example, Bastian—an opponent of the germ theory—cited Burdon Sanderson and Lister as its champions but then quoted from Burdon
Sanderson's experimental results to support his own positions.

In the debates of the mid-1870s Burdon Sanderson's position was not
unique. Confusion about how terms should be defined and about the germ
theory itself meant that many people found themselves on both sides of the
issue—at least according to others. Burdon Sanderson found himself a reluctant participant in the physicist John Tyndall's (1820–93) very public dispute with Bastian over spontaneous generation and the germ theory. Burdon
Sanderson was forced to enter the debate in order to explain his own, very
subtle position and how it had been distorted in summaries of his research
and remarks. Tyndall and Burdon Sanderson corresponded cordially about
the issue of the germ theory and upcoming debates in the Royal Society. (For
less complimentary private remarks about Burdon Sanderson by Tyndall see
the discussion in chap. 6.)[35] Bastian himself had named Beale a supporter of
a germ theory, when in fact Beale was opposed to the germ theory that bacteria caused disease. He had his own theory that disease germs were caused by
particles of degraded protoplasm derived from the diseased animal's protoplasm.[36]

To onlookers opinions and allegiances were even more difficult to discern
because of the overlapping controversy over spontaneous generation. For the
Contemporary Review the germ theory was the hypothesis that "no life has ever
been evolved (except in the remotest periods of the earth's history) otherwise than from a living parent or a living germ," as opposed to "the sponta-

neous generation theory," that "life does also spring *de novo* from molecular rearrangements of the atoms of dead organic materials."[37] Many historians have noted that the history of the emergence of a new concept of contagion resembled that of spontaneous generation. Moreover, the spontaneous generation controversy was also related to discussions of Darwin's theory of evolution. The resulting web of personal, professional, and philosophical allegiances was difficult to disentangle.[38]

In the British controversy Tyndall was a vocal supporter of Pasteur and thus an opponent of spontaneous generation. Somewhat ironically, Bastian, once a promising junior member of the Darwinian circle around Huxley, had come to espouse spontaneous generation because of his commitment to evolution. The association of the theory with materialism, radical politics, and amateur science led Huxley and other members of the X club—an exclusive dining club in which its members, including Huxley and Tyndall, discussed science—to dissociate themselves from it publicly. When Bastian not only refused to follow the line suggested by Huxley—that life had once originated but only in the distant past—but proceeded to engage in a protracted and public dispute with Tyndall and Pasteur himself, he found himself estranged from the Huxley circle. His warmer reception in the pathological community revealed that there was not a single, unified group that supported experimentalism in the life sciences in Britain; revealingly, no members of the X club were active medical practitioners.[39]

Burdon Sanderson refused to take a decisive position; in the same *Contemporary Review* article he was dubbed "the mysterious Dr. Sanderson, whom we dare not class either among the orthodox or the unbelievers." But he fueled the controversy when he testified that he had confirmed some of Bastian's experimental results. Burdon Sanderson's unimpeachable reputation as an experimentalist made it impossible for Bastian's critics simply to dismiss his findings as the result of sloppy techniques.

Despite the *Contemporary Review* article and some attention in the *Times,* the germ theory controversy was of less interest to a nonmedical audience than the debates about the causes of cattle plague had been. Cattle, unlike germs, had obvious economic importance. The confusion with the spontaneous generation debate heightened attention to the germ theory, yet, even in medical circles the germ theory as an etiological explanation generated relatively little interest; the controversy raged mainly within the small pathological community. This does not meant that there was no interest in the germ theory among medical practitioners and the public at large; germs were featured in both the medical and lay press. But the sheer amount of comment generated by the cattle plague epidemic was far greater. It is only in retrospect that the emergence of the germ theory gained greater prominence.

As Simon noted incisively in 1877, the lack of consensus over the meaning of specific terms such as *germ* or *organism,* let alone general terms like *life,* made the debates over the germ theory "as interminable as those which the schoolmen held six centuries ago about the dancing-power of Angels upon needle-points." Simon also chastised Bastian and Tyndall for arguing "too positively on the remote inferences from their experiments."[40]

Reacting to this controversy, Burdon Sanderson wrote, "From the way in which many intelligent persons talk, you might be led to suppose that they imagined that the question of infection might be removed from the field of pathology to that of meteorology." He added, "To those who have had to meet the difficulties which beset the path of the pathological investigator at every step . . . it appears so plain that the only way to gain knowledge about disease is the direct observation, experimental or clinical, of diseased processes."[41] Contemporaries were unwilling to accept Sanderson's opinion that the verdict was not yet in on the germ theory. But they heeded his call to look to direct observation for answers. And it was the laboratory rather than the clinic which was given the decisive role. Experiments played a pivotal role, with all sides quoting experimental results—indeed, often the same results to support their differing positions. During the discussion at the London Pathological Society, for example, both Bastian and Burdon Sanderson had underpinned their arguments by citing the results of their own experiments and the experiments of others. Their own research was so well known that another participant in the debate, John Dougall, simply stated, "Now, doubtless you all studied the details of Drs. Bastian and Sanderson's ingenious experiments on this question, and with your leave I shall trouble you with a brief account of two of my own." Dougall then explained how he had demonstrated that bacteria could thrive only in neutral, fairly alkaline, or faintly acid environments. When he added acid to blood serum in a test tube, the solution remained clear and odorless, while unadulterated blood serum "soon swarms with bacteria."[42]

The germ theories debates of the mid-1870s signified the ascendance of laboratory research over clinical observation in the development of etio-logical theories. Previously, the proponents of germ theories tended to be, as Drysdale put it in his 1878 review article, "natural historians and physi-cists" rather than medical practitioners. But Drysdale concluded that, more recently, prominent British pathologists had "pronounced in favour of the parasitic-germ theory."[43] Some of these pathologists straddled the bound-ary between researchers and practitioners, but the content of the discussions of germ theories which were published—in the *Lancet,* the *British Medical Journal,* and even the *Times,* for example—illustrated that confidence in the results of experiments was both widespread and not restricted to those ac-tively engaged in research. This belief in experimentalism I ascribe partly to the relative facility with which one could experiment on germs in this era (see Dougall's earlier remark), as opposed to carrying out vivisection ex-periments. With inexpensive, widely available equipment and little manual dexterity, anyone could attempt some germ experiments, and many did. There were, of course, technical problems—related to contamination, for in-stance—but the potential difficulties of such research were dwarfed by the problem of obtaining materials and equipment and learning the required techniques of even the most elementary vivisection experiments.

But these debates were about more than experiments. Several powerful personalities were also struggling to capture territory for themselves and their disciplines. Beale had been irritated with Burdon Sanderson and Simon since at least 1869. In a report on contagion for the Medical Office of the Privy Council, Burdon Sanderson had dismissed Beale's conclusion that the masses of germinal matter he had seen in vaccine lymph contained the con-tagious principle. Burdon Sanderson had written, "In the papers already re-ferred to, Dr. Beale expressed his opinion that the contagious principle re-sided in the granules, but he did not, to the best of my knowledge, offer any experimental proof of its being so."[44] Simon had committed the more un-pardonable sin of ignoring Beale's work. In response, Beale attacked Simon, writing to Burdon Sanderson: "When Simon conceived the tubercle & can-cer zymotic idea some wicked little microzymes must have been pirouetting amongst the particles of the living matter of the cells in the cortical portion of his cerebral convolutions. It is to be hoped that before long the more exalted bioplasm will again exert its sway & drive the microzymes into the sewage."[45] Beale sensed that the new experimental pathologists were conspiring to reject him as a microscopist of the old school. He was right to feel slighted; Burdon Sanderson's frequent references to morphologists, whom he castigated for simply cataloging bacteria rather than fighting disease, like pathologists, while not personal attacks, were directed at Beale's type of science.

The entry of Tyndall, formerly a physicist and famous polemicist, into the field of microscopic life also fueled the disputes.[46] Medical men in general resented what they perceived as Tyndall's ill-informed contributions to the discussions on the causes of disease.[47] Typically, in 1870 Beale had attempted to forge an alliance with Burdon Sanderson against the intrusion of Tyndall, writing, "Tyndall may dance on the tight rope & put the blue sky into his snuff box, but surely we who work at living things need not suggest that things *may differ* indefinitely as regards particulars which lie *beyond* the reach of microscopic investigation."[48] To further complicate matters, Beale was Tyndall's ally against Bastian in the row over spontaneous generation.

Tyndall himself was publicly polite to Burdon Sanderson, but he was usually quite dismissive of his critics. Huxley had written sympathetically to Tyndall: "Bastian has been blundering again as usual. There is an action in the Scottish courts 'for putting to silence' a frivolous litigant. I begin to wish we had something of the same sort for the irrepressible B."[49] By this time Burdon Sanderson was protected by his reputation. He was Britain's most prominent experimental pathologist, internationally recognized for having conducted experiments that had established that water and moist substances contained bacteria.[50]

Another undercurrent in these disputes was British nationalism. The debate about germs occurred at a time when Britons were becoming aware of external threats to their industrial strength. Following the triumphs of the Great Exhibition of 1851 and the London Exhibition of 1862, for example, it was evident at the 1867 Paris Exhibition that British products in many categories could no longer compete against those produced in foreign countries.[51]

British pathologists were resentful of the importance of French and German research to the establishment of the germ theory and resisted the dominance of their Continental colleagues. It was not unusual, during the discussion on the germ theory at the London Pathological Society, for Dougall to emphasize that "the painstaking and delicate investigations of Bastian, Sanderson and Beale into the life history of minute organisms . . . however they may differ in their conclusions do honour alike to *British scientific literature and to British experimental biology.*"[52] In 1875 Beale himself had publicly accused Burdon Sanderson of "disposing of the scientific work and conclusions of other Englishmen," among whose number Beale put himself.[53] Burdon Sanderson denied the accusation, but it was clear that he was looking to France and Germany for his models, not to British microscopists. Despite British researchers' attempts to participate in the international debate over the germ theory, it was already evident that the germ theories had originated on the Continent; French or German researchers were expected to solve the unanswered questions associated with the much-anticipated germ theory (as they, in fact, did). By 1883 the veterinarian James Lambert could lament that it was striking "how rarely any British names are mentioned" when reading about the researches connected with the germ theory; he acknowledged that British discoveries were "dwarfed by the greater ones of the Continental investigators."[54] The British were particularly annoyed because Jenner's triumph in establishing the vaccine against smallpox had previously made the idea of contagion appear to be Britain's property.

These debates also revealed the power of germ theories and the hope for a single explanation. There were still fundamental problems that everyone acknowledged—in particular, causality and specificity. No investigator had demonstrated that there was a causal relationship between microorganisms and disease processes, although even by 1875 it had been proven likely. T. Maclagan (who had also participated in the Pathological Society debate) wrote in 1876, "The balance of evidence being in favour of the view which regards contagium as consisting of minute organisms, we shall for the present assume that such is their real nature, and shall proceed to investigate the competence of such organisms to cause the phenomena of disease." At this point no investigator had been able to differentiate the microorganisms in order to attach a specific organism to a specific disease or to establish which were not pathogenic. As Maclagan noted in his 1876 essay, "All the bacteria which are seen in contagious fluids, and in diseased tissues, are not contagium particles."[55]

Contemporary experiments had produced much contradictory evidence. Some used such objections to promote the abandonment of germ theories in favor of the so-called physico-chemical theory that Bastian espoused, in which "contagia . . . [were] dead organic particles from altered-tissue elements or complex chemical compounds of alkaloidal constitution engendered in some of the tissues or fluids of the body."[56] The general consensus around 1875 appeared to be that the evidence was not yet convincing and that the experimental results and patterns of disease would in the end be explained not by a unicausal mechanism but one that integrated the influence of the environment (external and internal) and the microorganism.

Even in the face of such contradictory evidence considerable support for a germ theory had already emerged by 1875. As Bastian had remarked in 1872, "Like homœpathy and phrenology, this theory carried with it a kind of simplicity and attractiveness, which insured its acceptability in the minds of many."[57] In the first place the hoped-for germ theory had tremendous explanatory power—a huge number of phenomena, from cholera epidemics to puerperal fever to hospitalism, could be explained by one apparently simple mechanism, the action of parasitic microorganisms. The germ theory was almost a general theory of gravitation for pathology.

The attempt to use one theory to explain so many clinical phenomena, however, also fueled the controversy. So many disease processes were involved that complete closure of the debate in favor (or not) of the germ theory was impossible to attain. Thus, in an 1877 article Burdon Sanderson could applaud Koch for his elucidation of the life cycle of the anthrax bacillus—Koch's 1876 publication is traditionally cited as the turning point in the acceptance of the germ theory—and simultaneously reject germs as the cause of septicemia.[58]

In the mid-1870s the second great appeal of a germ theory was its simple mechanism, which gave practitioners hope that effective therapies or preventive techniques would quickly follow. In part medical practitioners reflected the optimistic spirit of mid-Victorian Britain; their optimism was also underpinned by the belief that current (i.e., pre–germ theory) sanitary practices had already significantly improved the health of Britons. They expected that the germ theory in its turn would lead immediately to further improvements. In the mid-1870s the fact that these organisms could be destroyed by heat and disinfectants and the claim that the germ theory had already resulted in the successes of the Listerian surgical techniques reinforced the general sense of optimism. By the early 1890s, however, this optimism was seen to have been premature.[59]

The late 1870s were a very difficult time for Burdon Sanderson. His 1875 study on inflammation was the last research he carried out for the Privy Council. He had been frequently ill and reluctantly declined any further commissions, writing to Simon that his "ordinary professional occupation" was, in his present state of health, "more than sufficient."[60] In 1876 a family tragedy thrust more responsibility on Burdon Sanderson. His elder brother Richard's family was involved in a serious railway accident. Richard had married Isabella Michelson Haldane, the sister of Robert Haldane—their sister Mary's husband. The accident killed his two nieces immediately and seriously injured Isabella and his brother, who only survived for three months.

John's own health did not improve, and he was often despondent. He wrote in his diary at the end of 1878: "Feel so depressed—so tired. Dread my lectures—a cloud comes over me as I go into the lecture room. My head aches." Later he noted that during the week "all the time the depression seemed only just below the surface, and ready to break out at any moment." His mood was very noticeable, and he snapped at Ghetal when she "said in fun 'Bring your thoughts down'—'Bring them down, as if they were not as low as they could be already!'"[61]

Burdon Sanderson's melancholy outlook makes it difficult to distinguish whether he was depressed because he was ill or whether his depression was his illness. He had always worked extremely long hours without ever feeling that he had accomplished enough; any illness was likely to make him despondent. Indeed, his grueling work schedule, as his own physicians believed, must have contributed to his physical decline. In 1878, at the age of forty-nine, Burdon Sanderson agreed to reduce his workload. Under the circumstances he resigned as professor superintendent of the Brown Institution and resolved to give up his pathological research in favor of physiological research. Burdon Sanderson chose physiology over pathology partly because of his own personal interest. His decision also revealed his opinion, articulated on other occasions, that physiology ranked above pathology. Nonetheless, there is every reason to believe that, had ill health not intervened, Burdon Sanderson would have remained active as a research pathologist.

CHAPTER SIX

Focusing on Physiology

Capturing the Venus's-Flytrap's
Electrical Activity

Men sneered at vivisection, and yet look at its results today! . . .
did I hold the key to the fancy of even one lunatic—I might
advance my own branch of science to a pitch compared with
which Burdon-Sanderson's physiology or Ferrier's
brain-knowledge would be as nothing.   Bram Stoker, *Dracula*

In the early 1870s, when Charles Darwin (1809–82) turned again to a project that had captured his interest in the 1860s—studying carnivorous
plants—he was referred to John Burdon Sanderson for some of his more
physiological questions. In the course of their correspondence Darwin suggested to Burdon Sanderson that he investigate whether there was any electrical change in the leaves of *Drosera* or *Dionæa muscipula* (the Venus's-flytrap)
when they were excited.[1] Darwin lent Burdon Sanderson "two plants with
5 goodish leaves" on 9 September 1873 and wrote in closing: "Whenever you
have quite finished I will send for the plants in their basket. My son Frank is
staying at 6 Queen Anne St, & comes home on Sat. afternoon, but you will
not have finished by that time."[2] That month Burdon Sanderson announced
that he had discovered that when the leaf of a Venus's-flytrap was stimulated
manually, a voltaic current was generated in the leaf before it moved to close
on its supposed victim. The currents of the flytrap leaf were, he concluded,
"subject, in all respects in which they have been as yet investigated, to the
same laws as those of muscle and nerve."[3]

These results so captivated Burdon Sanderson that the electrical phenomena of the Venus's-flytrap became the focus of his physiological research

and, when he gave up pathological research in 1878, of his entire research agenda. The reception of these experiments led him to examine the nature of the excitatory process in the comparable animal structures.[4] Until 1888 Burdon Sanderson usually spent the summers doing research on the Venus's-flytraps, first at Kew Gardens and then at his own laboratory either at University College or Oxford—where he moved in 1883 to become the first Waynflete Professor of Physiology. In the winters he carried out related animal experiments. He believed that he had established the existence of leaf currents like those found in animals.[5] Thus, his experiments belonged to the long-standing tradition of searching out analogies between animals and plants.[6]

Burdon Sanderson's use of the Venus's-flytrap, a carnivorous plant with menacing qualities, made these experiments even more compelling, particularly for a popular audience. He twice used them in lectures at the Royal Institution which featured elaborate demonstrations of excitability. The demystification of a familiar monster heightened the impact of the flytrap experiments. In these popular lectures he seized the opportunity to explain the importance of the experimental method for physiology and to connect contemporary physiology to the old observing sciences of life practiced by anatomists and naturalists.

The Venus's-flytrap work illustrates Burdon Sanderson's ability as a detail-oriented experimenter—someone who built his own instruments and was caught up in the fine details of his research. In contrast to Michael Foster (1836–1907), who was not known as a particularly able or active experimenter, Burdon Sanderson was a hands-on resource for his colleagues. Burdon Sanderson's influence was far-reaching and cut across disciplinary boundaries (see the appendix, which lists the many researchers who spent time in his orbit).

The Venus's-flytrap experiments also epitomize the communal, cooperative, and international nature of experimental practice at this time. The rheotome and capillary electrometer, two complex instruments, reveal how materials and techniques were imported from the Continent and altered to suit Burdon Sanderson's own experiments.

After his medical training Burdon Sanderson had entered French research circles: he had studied with Bernard in 1851–52, and in his pathological research he had worked with Chauveau. He never cut his ties to France: in the early 1870s, as he worked on the *Handbook for the Physiological Labora-tory,* he traveled to Paris to consult Bernard, among others. Nonetheless, the overwhelming importance of German-language researchers, their tech-niques and results, meant that Burdon Sanderson turned his attention in-creasingly toward this different group of researchers, to such an extent that he was apparently unaware that Bernard was also engaged on experiments in plant physiology before his death in 1878.[7] Burdon Sanderson's close ties to German research circles, illustrated here through his adoption of specific instruments and methods, contributed to his utility to his British colleagues.

With his Venus's-flytrap research Burdon Sanderson attempted to estab-lish himself as an experimental physiologist in the international community, but that was not his sole motivation: his fondest wish was to contribute to the sum of total knowledge about nature. On many public and private occasions Burdon Sanderson expressed his belief in the importance of making a perma-nent contribution to science, writing in 1873: "In pathology just as in physi-ology—in physiology just as in physics and chemistry—the criterion of value is not utility but permanency. A little bit of work done once for good and all, is what each scientific aspirant ought to aim at."[8] In his scientific papers Burdon Sanderson aimed to gain acceptance for his discovery of a "normal leaf cur-rent" and the "negative variation" in that current which occurred after either mechanical or electrical excitation of the leaf. His research on the leaf of the Venus's-flytrap was, for Burdon Sanderson, the pivotal accomplishment of his career, the summit of his achievement.

Burdon Sanderson did gain acceptance within the international physio-logical community, although a research school did not grow out of his work, at least not one accepted by contemporaries. I found only two references to an "Oxford School of Physiology"—in contrast to the ubiquitous refer-ences to the Cambridge School of Physiology. In 1895 George Buckmaster, a former pupil, congratulated Burdon Sanderson on his new post as Regius Professor of Medicine. Buckmaster hoped, however, that Burdon Sander-son would not lose interest in physiology—"especially," he explained, "as the school at Oxford was founded by you and the present position of the science in this country is largely due to your work and influence." A year later E. A. Schäfer wrote to ask Burdon Sanderson if he, "in conjunction with your col-leagues of the Oxford School," would write the section on the physics of nerve and muscle for a new textbook of physiology.[9] Burdon Sanderson established himself as a specialist in the field of electrophysiology, but he was unable to gather a core of researchers around him at Oxford.

This failure was also due to the philosophical milieu. In late Victorian Britain, Evangelicalism had been to some extent superseded by utilitarian, positivist, or idealist beliefs. The Evangelical impulse had not disappeared; like Burdon Sanderson, many developed a personal philosophy that incorporated elements of different philosophical approaches. Some reacted to Evangelicalism; others attempted to combine Evangelical religion with their new beliefs. But by the late Victoria period Burdon Sanderson's brand of empiricism appeared outmoded and, more significantly, made it difficult for his junior colleagues to integrate their research into the larger questions being examined in the field.

In this period of his life Burdon Sanderson left medical practice and pathological research behind him. His research, with its focus on a carnivorous plant, was at first glance entirely divorced from medicine. But there were continuities between this work and his earlier work on circulation and respiration. Both aimed to register and analyze the phenomena of life. The importance of electrical activity to nervous activity in the animal body had long been established. In a material sense these experiments derived from earlier animal experiments and utilized many of the instruments that Burdon Sanderson had developed or refined in conducting that research. In addition his interest in finding "disinterested observers" who would produce unequivocal results that anyone could directly read continued, and he produced graphical records of the Venus's-flytrap's electrical activity which resembled his sphygmographic tracings.

In the 1880s Burdon Sanderson focused on his physiological research. He aimed to make discoveries and to form a research school in the field. The representations of the electrical activity of the Venus's-flytrap and its animal analogues, which he produced, are emblematic of Burdon Sanderson's vision of experimental physiology.[10] His representations also revealed the limitations of his vision; in the end all that Burdon Sanderson possessed were representations. He made no real attempt to explain the phenomena underlying his representations. His hesitancy to make what he considered unwarranted speculations was no longer acceptable in British scientific circles, in which idealism was on the ascendant, or abroad. Burdon Sanderson's empiricism— his belief that science merely made evident facts—contributed both to his inability to establish a research school that would rival that of his contemporary, Foster, of Cambridge, and to the reluctance of others to accept his discovery of the electrical activity of the flytrap.

The emerging divisions among the life sciences also played a role in the reception of Burdon Sanderson's Venus's-flytrap research. By the end of the nineteenth century, even within medical schools, botanists and comparative anatomists or morphologists felt threatened by the importance of physiology. Outside the medical milieu men like Huxley and Foster aimed to create a pure science of biology which did not depend on medical utility. Burdon Sanderson's Venus's-flytrap research did not fit neatly into a single disciplinary box and therefore did not find a ready audience.

Burdon Sanderson's Electrophysiological Research

When a fly (or other unfortunate creature) touched one of the Venus's-flytrap's sensitive hairs, the leaf would quickly close by folding itself along its midrib. The three sensitive hairs and the midrib are clearly visible in figure 6.1 (to the left and below the letter *a*).[11] The Venus's-flytrap itself, literally a sensitive plant—difficult to grow and easily killed—was an important limiting factor for experiments. Burdon Sanderson could only obtain leaves of the necessary size in the summertime at Kew Gardens. In addition, Burdon Sanderson's technique destroyed each leaf that he experimented upon.

For his first experiments Burdon Sanderson created simple circuits, using Emil DuBois-Reymond's (1818–96) electrodes (see fig. 6.2), the leaf of *Dionæa*, and a Thomson galvanometer. He found that a resting leaf had two opposing currents, the "normal leaf-current" and the "normal stalk current." Here, although Burdon Sanderson credited DuBois-Reymond explicitly only for the electrodes and the induction apparatus, he was closely influenced by the researcher's 1840s work on animals; he also deliberately borrowed DuBois-Reymond's terminology. By the 1870s these techniques were in standard use in the field of animal electrophysiology. Burdon Sanderson's use of the Thomson galvanometer was innovative.[12]

Burdon Sanderson illustrated the arrangement of the leaf on the electrodes—two arrows indicated the direction of the leaf current and, in the opposite direction, the stalk current. Any stimulation of the hairs of the leaf or the leaf itself, whether electrical or mechanical, Burdon Sanderson found resulted in a deflection of the galvanometer needle away from the direction of the normal leaf current and in the direction of the normal stalk current. This was the "negative variation,"[13] which he contended mirrored DuBois-Reymond's earlier discovery of the same phenomenon in animal muscle. In his textbook Ludimar Hermann (1838–1914) described how, "when the whole of a muscle whose external surface and artificial cross section are connected with the galvanometer is thrown into contraction, there is a diminution of the muscular current, a '*negative variation*' (DuBois-Reymond)."[14]

In May 1876 Burdon Sanderson went to an exhibition of scientific appa-

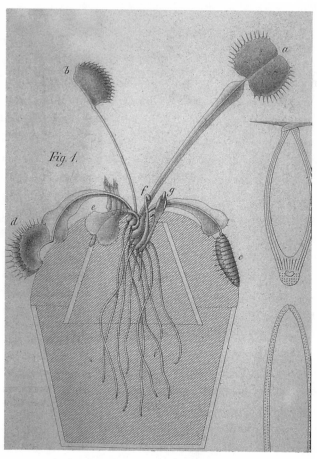

Figure 6.1. The Venus's-flytrap,
Yale University Biomedical Communications.

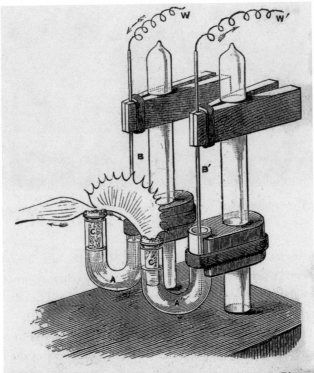

FIG. 2.—Non-polarisable electrodes employed in experiments on the Dionæa leaf.

Figure 6.2. The flytrap leaf preparation,
Yale University Biomedical Communications.

ratuses in London. There he saw the capillary electrometer, which had been developed by G. Lippmann and Marey (see fig. 6.3).[15] It consisted of a small column of mercury in a bath of dilute sulfuric acid with a potential applied across the mercury/sulfuric boundary: any change in potential would lead the column of mercury to rise or fall. Using a capillary electrometer, an investigator could see, and measure, very small changes in voltage.[16] Burdon Sanderson realized that such an instrument would be useful for his flytrap experiments.[17]

In order to reinforce his discovery of the unity of the electrical phenomena of the Venus's-flytrap with that of animals Burdon Sanderson continued to explore the phenomena of animal excitation (in collaboration with his assistant, F. J. M Page [1848–1907]). These experiments have been identified as a

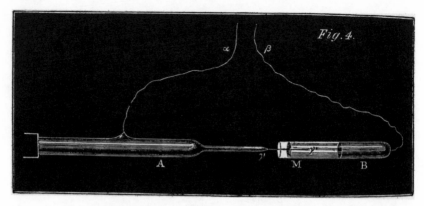

Figure 6.3. The capillary electrometer, Yale University Biomedical Communications.

step in the understanding of the electrical nature of the heartbeat.[18] He produced several papers on the time relations of the excitatory process in the ventricle of the frog,[19] and in two 1880 publications he described his experiments on the frog heart in detail.[20]

With the results of their technically difficult experiments, Burdon Sanderson and Page aimed to describe more precisely the nature of the electrical changes in the frog ventricle during excitation. The complexity of their task was illustrated by their experimental setup (see fig. 6.4):[21] "when the keys K, K1, K2, K3, and K5, are open, and the switch W is in the position shown, any current between f and m passes through the galvanometer G, the compensator C, and the rheotome R. Only one of the four pairs of mercury pools of the rheotome is used, so that the circuit is closed once in each revolution."[22] Burdon Sanderson and Page began their study of the electrical variation using the capillary electrometer. Immediately after exciting the leaf, they observed the mercury column move downward, followed after a pause by the column moving upward. They could measure the movement of the mercury column in "degrees of the electrometer scale," but for a more quantitative picture they changed technique. Burdon Sanderson and Page used a rheotome, designed by Burdon Sanderson himself (see fig. 6.5), which allowed an observer to take galvanometer readings at fixed intervals after the excitation. His design epitomized his hands-on approach to experimentation.

A rheotome consisted of two independent circuits, one the exciting and the other the measuring, or "galvanoscopic," circuit. The two circuits were closed and opened in succession. During experiments the investigator needed to be able to vary, and to measure accurately, the time interval between the two closures and the period of closure of the galvanoscopic circuit. There were two rheotomes in common use at the time, both developed by men who had worked with Dubois-Reymond. With Hermann's instrument an experimenter determined time intervals by measuring the distance a falling weight passed through in the interval between the first and second event. Using the design of Julius Bernstein (1839–1917), a close friend of Hermann, who had also spent four years with Hermann Helmholtz (1821–94) in Heidelberg, the investigator measured the intervals by the rotation time of a wheel driven by an electromagnetic motor.[23]

Burdon Sanderson's ingenious design consisted of a segmented cylinder (see fig. 6.5). In the segments were pools of mercury connected by conductors. As the cylinder was rotated, the points of contact were also rotated, and thus two different circuits were opened and closed. The period could be changed by varying the distance of the contact screw, *f*, from the center of the cylinder. The instrument in figure 6.5 was drawn as arranged for the frog ventricle experiments.

With this experimental setup Burdon Sanderson and Page analyzed in detail the nature of the excitatory variation in the frog heart. They studied how the variation changed with temperature and injury to the heart and how it varied across the heart. The unexcited parts of the surface of the ventricle were "*negative*" to every unexcited part, during "the excitatory state"; this state was propagated in all directions at a rate of about 125 millimeters per second (measuring from apex to base). "The effect of partially warming

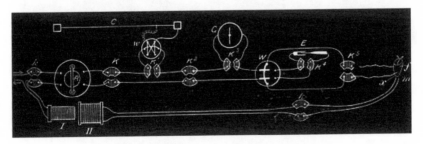

Figure 6.4. Schematic of instruments in frog experiments,
Queen's Medical Photography.

the surface at the ventricle," they wrote, "is to shorten the local duration of the excitatory state at the part warmed." The effect of a localized injury "is to diminish or weaken the excitatory state (Reizwelle) at the injured part."[24]

At this point Burdon Sanderson's beliefs about the relationship between the animal and plant phenomena had undergone a subtle change. Initially, he had stated that plant excitation resembled that in the muscles and nerves of the body—as demonstrated, for example, in DuBois-Reymond's classic frog *gastrocnemius* muscle preparation. Now, incorporating the results of his own experiments, he stated that plant excitation was like that in analogous animal organs, such as amphibian hearts.

In the late 1870s and early 1880s Burdon Sanderson returned to electrical plants in the wake of German work that challenged his discovery. Professor Hermann Munk (1839–1912) of Berlin had contradicted Burdon Sanderson in 1876. A year later A. J. Kunkel, a plant physiologist, had concluded that "all electromotive phenomena observed in the organs of plants are dependent on

Figure 6.5. Rheotome of Burdon Sanderson's design,
Queen's Medical Photography.

changes in the distribution of water in their tissues, and consequently have nothing whatever in common with the electromotive phenomena of muscle and nerve."[25]

In 1881 Burdon Sanderson published a long paper on the flytrap which included a critique of Munk's 1876 paper. Most striking were their two very different experimental styles. Munk's conclusions were rooted in careful anatomical studies carried out by Fritz Kurtz (fig. 6.1 is from Kurtz's paper). Munk experimented on hundreds of leaves, and he did not detach the leaves from the plant or fix them open, as Burdon Sanderson did. Munk's conclusions were embedded in theoretical explanations of the electrical activity, which Burdon Sanderson deplored. Munk concluded that one could replicate the electrical properties of the leaf by aligning zinc cylinders with copper zones, as the cells of leaf are aligned, and surrounding them with a uniform layer of moist conductor (this schema was similar to that of DuBois-Reymond's for elements of muscle, which he had compared to copper cylinders with zinc zones). Munk denied the existence of the "negative variation" or its connection with the mechanical effect of leave closure.

Munk severely chastised his English colleague for what Burdon Sanderson explained had been a "casual expression" in his 1874 Royal Institution lecture "to the effect that the excitatory motion of the leaf is identical with muscular motion." In addition he denied making such a comment in his scientific papers and explained that "I meant nothing more than that both motions are expressions of the same elementary endowment of protoplasm—that of changing its form on excitation. I held then, and hold now, that, as regards what happens in the living protoplasm, the mechanical effects of excitation in the contractile parts of plants & animals are closely related. As to the mechanism . . . nothing was said in my paper."[26] Burdon Sanderson did not directly address Munk's main point that the supposed similarities that could be found between animal and plant phenomena were not physiologically meaningful.

Burdon Sanderson's avoidance of explanation separated him from his German colleagues. German physiologists were engaged in long-standing debates about the possible electrical and molecular explanations of the electrical phenomena found in animals. It was these debates that fueled their experiments.[27]

Burdon Sanderson's Place in the International Community

Before this line of research, Burdon Sanderson had developed relationships with many Continental researchers. He was acquainted with most of the major research figures of the period. Through his and Ghetal's letters and diaries, a picture of him as a successful networker emerges. Burdon Sanderson's career illuminates how research success depended on personal contacts.

Burdon Sanderson sought introductions to other physiologists through his friends and research associates. He also bought the latest apparatus from foreign suppliers.[28] Most importantly Sanderson made brief visits abroad to make personal contact and work with many prominent investigators. And, of course, international figures visited London.[29]

In 1870 the Burdon Sandersons paid a visit to Ludwig at Leipzig. Ghetal, a tolerant wife (who had entirely recovered from her 'mild case' of smallpox) gave an account of their stay to her sister-in-law, Mary. She explained that their planned departure had been postponed because, "John got into the midst of experiments and what not and occupied himself therewith the whole of that day and half the next. Then on Tuesday afternoon he had several small purchases to make of materials for experimenting, so that we did not leave Leipzig until 6 p.m." Ghetal was confident that their plans to return to Leipzig the next day were quite definite, "as a basket of frogs is to meet us at the Leipzig station and to accompany us on our way home. I am not very enthusiastic about having such traveling companions but John declares they will be no trouble whatever." The Burdon Sandersons were not only traveling with frogs; Ghetal added, "We have already a long roll of paper almost as tall as myself, and as we are going to Jena to see more Professors I don't know what more articles we may collect before we arrive in England."[30]

Since Burdon Sanderson had teaching responsibilities in England, he could not make long trips to the continent. He maintained contact with letters, and through his junior colleagues who could make extended visits to foreign laboratories. During E. A. Schäfer's (1850–1935) trip abroad in 1872, for example, the two men corresponded frequently. Burdon Sanderson sent greetings to foreign colleagues, advised his younger colleague, obtained information about ongoing research projects abroad, and authorized Schäfer to buy instruments.[31] Throughout his career Burdon Sanderson continued this pattern, thus he maintained his own contacts and fostered the careers of young researchers.[32]

Burdon Sanderson believed that direct connections to experts were required in order to carry out research properly. Well aware that "by far the greater number of those who exercise themselves in scientific investigation fail to accomplish any permanent work at all," he blamed these failures in part on carelessness, and the reluctance of some investigators "to be guided in their work by the great leaders of scientific thought." He asserted, "A man who is minded to work at any particular subject has . . . to study carefully what other workers, and especially what the great leaders of scientific thought have stated about it."[33]

Burdon Sanderson must have been stung by a scathing letter which DuBois-Reymond wrote in response to his 1876 paper "Note on the Electromotive Properties of Muscle." DuBois-Reymond accused Burdon Sanderson of ignorance of his latest publications, writing, "I regret being obliged to tell you, that, as a friend of English science, I have been truly mortified at such want of knowledge and such rashness on the part of a man in your high scientific position. It would be easy for me to expose you publicly in a way unpleasant to you. I however prefer pointing out to you privately the errors into which you have fallen, and the means of correcting yourself."[34] He graciously offered to be Burdon Sanderson's "guide in the dangerous domain of Animal Electricity." Burdon Sanderson publicly retracted his comments.[35]

In the years that followed, Burdon Sanderson went to Europe frequently, where he visited other physiologists. His diaries and letters mention many visits to prominent colleagues and their laboratories. He wrote in October 1877, "In Breslau and Halle I accomplished all I wanted and was most cordially received by everyone."[36] In 1879 Ghetal noted that, "J[ohn] is gone to Zurich to see Prof. Hermann."[37] The following year, while in Berlin, Ghetal recorded,"Next morning we went to Utrecht where we saw Profs. Donders and Engelmann, who were so disappointed that John could not stay longer that we had to promise to go there again on our return." In Berlin, they met Hugo Kronecker (1839–1914) and called on the DuBois-Reymonds; at a dinner at Kronecker's home, Helmholtz and Julius Cohnheim (1839–84) were among the other guests.[38] On the same trip Burdon Sanderson wrote to Schäfer, "As you see we are on our way home . . . The day before [yesterday] we were at Leipzig & saw Ludwig, who I was sorry to find a good deal altered."[39]

During these trips, as Ghetal had so colorfully explained to her sister-in-law, Burdon Sanderson saw the latest techniques, bought instruments, and discussed problems in his own research. The Burdon Sandersons managed to fit in time for socializing and sight-seeing; their diaries are full of accounts of walking tours. As he became established as a research physiologist and better acquainted with his continental colleagues, the trips abroad grew even more sociable.

A concrete example of the importance of these trips to Burdon Sanderson's research was his adoption and modification of the capillary electrometer. He encountered the instrument in 1876. While working with it, he found it needed to be adjusted. He consulted with Fleische of Vienna, who replied that he would be gratified to have his instrument noted. He then apologized in turn for an error he had made in print about Burdon Sanderson's research.[40] Sanderson had also written to C. Scøven for details about his electrometer. Scøven promised to bring along an instrument of his construction on his upcoming visit to London.[41] After these consultations, Burdon Sanderson constructed an electrometer of his own design with Page. This was the instrument he used in further experiments on Dionæa and on the electromotive properties of muscles.

Burdon Sanderson's abilities as an experimentalist were established by his investigations into plant and animal excitability. He was considered an expert in electrophysiology, which was why Schäfer asked him write the sections on nerves for his new physiology textbook. By the 1880s Burdon Sanderson was firmly established as a member of the international community of physiologists. And after the chastising correspondence of 1877, he even became a translator and close friend of DuBois-Reymond.[42]

The Philosophical Context: Empiricism and Idealism

Burdon Sanderson once wrote, "The study of the life of plants and animals is in a very large measure an affair of measurement."[43] This statement summarized his empirical research style. Even more succinct was the title of the painting by J. S. Marks, *Science Is Measurement,* which Burdon Sanderson himself had inspired.[44]

The production of quantitative data—as opposed to qualitative—was what defined the "scientific study of nature." Burdon Sanderson explained that the process of science was comparison with known standards, and "the completeness of knowledge is to be judged of in the first place by the certainty of the standard which we use; and secondly, the accuracy of the modes of comparison which we employ." He continued, "comparison with a standard is simply another expression for measurement."

Here Burdon Sanderson addressed the biology section of an instrument exhibition; he aimed explicitly to separate biology from its descriptive, 'naturalist' past. He implicitly separated physiology and pathology—his first specific example was pathological, the rest were physiological—from other biological sciences like morphology which were more closely associated with the naturalist approach. Indeed his address illustrated that the boundaries between these "biological" fields remained blurred, although in retrospect the emerging distinctions are evident.

When Burdon Sanderson said that physiology should be built on physics and chemistry, as we have seen, he meant that physiology should use the instruments and measuring techniques of chemistry and physics, and should depend on those subjects for its underlying explanations of phenomena. Burdon Sanderson stated that work in the biological sciences differed in only one respect from that in the physical sciences, "we accept the standards [of the physical investigator] as already established, and are content to use them as our starting-point in the investigation of the phenomena which concern us."[45]

Burdon Sanderson's research program was shaped by his reaction to criticisms of his results, and driven by instrumentation. His instrumentalist approach was epitomized in the Venus's-flytrap experiments. In his experiments the leaf was divorced from its natural state and became a part of the instrument. The importance of the evolution of technique and instruments was for Burdon Sanderson paramount. He began with a galvanometer to measure the electrical activity, but such an instrument left little potential for further research. The capillary electrometer and the rheotome allowed for more detailed analysis of the phenomena. The photographic method, discussed below, fulfilled his primary desire—to reproduce the phenomena he had first seen in 1873.

Burdon Sanderson was engrossed by instruments also because of his empiricist stance. Instruments produced tracings and then photographs he could measure; instruments had dials to read. All of his publications demonstrated that he would rather discuss results than attempt to explain them; he avoided discussions that moved much beyond his data. Burdon Sanderson—echoing some of his French predecessors like Laennec—considered such discussions to be unacceptable theorizing, to be unscientific—metaphysical. His personality played a role here. By all accounts, Burdon Sanderson became increasingly withdrawn as he got older, almost timid: perhaps to create a bold theory or to attempt to synthesize the results of many experiments required an audacity that he lacked.

In addition there were strong philosophical precedents for Burdon Sanderson's stance, deriving from a long-standing tradition of British empiricist philosophers—most notably Francis Bacon (1561–1626) and David Hume (1711–76). There was also a medical/empirical tradition, evident, for example in the Parisian medicine of the early nineteenth century. As I discussed in chapter one, although Comte himself was from an earlier era, his positivism was still very influential among contemporary scientific naturalists of Burdon Sanderson's era.[46] The philosopher Ian Hacking has written, "Untestable propositions, unobservable entities, causes, deep explanation—these, says the positivist, are the stuff of metaphysics and must be put behind us."[47]

Not all Victorians rejected metaphysics and rejoiced in the simple collection of data. Another John Scott, J. S. Haldane (1860–1936), Burdon Sanderson's nephew and colleague at Oxford, revealed their philosophical differences, when he commented in 1891, "[Burdon Sanderson] would say, I dare say, that he is very tolerant about theories—that what really tells is facts. But then what are the facts that are essential? It's the theory that determines that. I would simply disregard as trivial & misleading heaps of things which he considers essential, & vice versa. And even the simplest 'facts' are expressed, perceived through theory."[48] Haldane embraced an idealistic worldview. "The idealists," as Sandra den Otter noted, "adopted elements of both Kantian and Hegelian philosophy and language, prompting one contemporary to characterize the idealist school as a 'parrot-like imitation of German cloud-cuckoo-land.'"[49] Often believers in an expanded role for the state, idealists sought a role for the moral and the social good in the affairs of the world; in other words, most idealists opposed the laissez-faire political philosophy of classical liberalism. They rejected empiricism, in part, because of the connection of scientific naturalism—most evident in the writings of Herbert Spencer(1820–1903)—"to a vindication of Utilitarianism and laissez faire."[50] Although ostensibly idealism was separate from empiricism, in practice, as Burdon Sanderson illustrated, the lines were often blurred. So-called empiricists employed ideas or principles which were central to idealism—most notably the common good and community, and similarly idealists claimed to uphold certain empiricist convictions, but not with any materialist or naturalist baggage.

Burdon Sanderson might have been expected to be sympathetic to idealism. He was not hostile to German views, and his beloved sister, Jenny, was engaged in a translation of Hegel's *Lectures on the Philosophy of Religion, together with a work on the proofs of the existence of God* before her death in 1889.[51] He evidently supported a role for the state; Simon, for example, had been strongly influenced by T. H. Green (1836–82), an leading exponent of idealism.[52] Through his nephew, the politician R. B. Haldane (and brother of J. S.), Burdon Sanderson was directly connected to the New Liberalism rooted in the idealist belief in the importance of whole societies, rather than simply the needs of individuals. In an 1901 address which illustrated his commitment to social reform, Burdon Sanderson spoke on "Our Duty to the Consumptive Bread-earner." When he evoked "common Christian charity," as well as the obligations of citizens to not waste a valuable commodity—"the earning power of the workman"—Burdon Sanderson drew on a combination of Christian belief and social activism to exhort his audience of medical practitioners to action.[53]

Burdon Sanderson usually compartmentalized his life. He did not, in fact, entirely reject speculative thought, after all he was still a Christian (differently from the Haldanes, who moved from the Evangelicalism of their youths to idealism). Nonetheless, as a working scientist, he remained an empiricist.

The Failure to Establish a School of Physiology

Burdon Sanderson pursued his career in the shadow of the Cambridge School of Physiology. Throughout his life he was disappointed by his failure to create a similar school, particularly at Oxford in the 1880s.[54] Both his personality and his research style were obstacles, but other factors played a role. In the next chapter we shall see how the politics of the University of Oxford contributed to the failure of the Oxford physiology program.

Burdon Sanderson's positive influence on his contemporaries has been underrated. Setting this negative tone and reflecting the opinion of Burdon Sanderson in the Huxley-Foster circle, Schäfer, in his *History of the Physiology Society,* noted that, "although Sanderson was himself a strenuous worker, he appeared not to have the same power of instigating research as his Cambridge colleague." There were "a certain number of men who afterwards attained distinction worked under him, but their number was insignificant in comparison with those attracted by Foster at Cambridge."[55]

When his colleague at Oxford, E. Ray Lankester (1847–1929), noted, "I agree fully with your suggestion about the sea-water experiment: and am very glad that you encourage Vernon to go into the matter," he revealed Burdon Sanderson's regular practice of assisting his more promising young students.[56] At Oxford Burdon Sanderson also supported the careers of his junior colleagues James Ritchie (1836–82) and Wynstan Dunstan. After his appointment as professor of pathology at Oxford, Ritchie wrote, "Without your initiative and constant encouragement this new and more responsible appointment would never have come into existence." Dunstan taught pharmacological subjects at Oxford and was supported by Burdon Sanderson as he progressed through the ranks.[57]

Above all, Burdon Sanderson was recognized as a first-class experimenter; his correspondence contained many requests from British (and foreign) investigators for specific and general advice about their own research. These men did not form a Burdon Sanderson school, and he was not solely responsible for their research achievements. But he devoted large amounts of his time to reading and critiquing the articles of younger colleagues; he visited their laboratories, and they visited his. Such activities continued until the end of his life. In particular Walter Holbrook Gaskell (1847–1914) and George

John Romanes (1848–94), two of the most prominent members of the Foster School of Physiology, spent several years working with Burdon Sanderson in London and remained his lifelong friends.

Burdon Sanderson's foreign recognition also enhanced the development of British physiology. His connections brought E. E. Klein (1844–1925) to London. In 1871 he hired Klein, who went on to a career as a research pathologist and bacteriologist, on the recommendation of Professor Stricker of Vienna.[58]

Burdon Sanderson also tried to protect and promote junior colleagues. In the 1870s he attempted, unsuccessfully, to secure a University College London professorship for Schäfer. An important factor in Burdon Sanderson's decision to leave University College for Oxford was his sense that he was standing in the way of Schäfer's well-deserved promotion. In December 1878 Ghetal recorded that John had returned from a trip to Cambridge "looking dismal," after he had "got on a bad train of thought on the way home." John had realized that his "position at U.C. will soon be untenable. S[chäfer] must always feel that I stand in his way."[59] Burdon Sanderson's involvement in Schäfer's research and career progress was greater than Schäfer ever acknowledged.[60]

Burdon Sanderson was most successful in promoting his protégé, Francis Gotch (1853–1913).[61] He worked behind the scenes to get Gotch into the Royal Society and made sure that Gotch followed him as Waynflete Professor at Oxford. Gotch wrote to Burdon Sanderson: "I attribute all my work to your influence and, if I may say so, to what I believe is very rare even in scientific work to your genial fellowship with a young and inexperienced worker. Now that I am away from Oxford I look back and value even still more highly than before the stimulus which impels others towards original investigation and patient experimental observation many times repeated which I know now to have really come from yourself."[62]

There were many other examples of his patronage of and influence on upcoming researchers throughout his career.[63] Charles Beevor wrote to Burdon Sanderson after Sanderson was made a baronet, "Personally I always felt most indebted to you for your lectures and teaching at University College which I am sure have had great influence on one's work in making one look at medicine from the true scientific point of view."[64] After getting a post, Victor Horsley (1857–1916) wrote to express "not only my thanks but also my sense of the fact that my only qualifications which have secured for me so honour-able a gift were inculcated by you in the Physiology Courses at Univ. Coll. Lond in 1875–6-7."[65] E. B. Titchener, Sage Professor of Psychology at Cornell, wrote: "I thought you would be glad to hear of my success; and I wished at the same time to express my deepest thanks to you for your kindness to me at Oxford. I cannot overestimate the advantage which I derived from my year in the Physiological Laboratory."[66] Fred Mott described himself in a letter to Sanderson as "an old pupil of yours and one who owes in great measure any success he may have attained to the spirit of scientific enquiry which you stimulated & fostered."[67]

Burdon Sanderson's relationships with his students were not ideal. His tendency to stand on precedence could be stultifying.[68] His belief in the necessity of consulting the experts, the leaders in the field, was an outgrowth of his hierarchical approach to life. Burdon Sanderson was also a tireless worker, who expected no less of his students. He could be pompous, but he was most notoriously absent-minded and timid. An anecdote (one of many) which illustrated both his absent-mindedness and his experimental sangfroid was related by John Abel: "One of Sewall's recent stories was about Prof. Burdon Sanderson, the famous London professor. He is a great, tall awkward man and one day while eating his dinner in the laboratory where he was experimenting with frogs, actually bit into one in place of the sandwich."[69]

A former student writing to Lady Burdon Sanderson about an old letter of his to her husband, stated, "What stands out most closely in the letter, and is more important than my own results is the absolute and explicit trust that I had in your Husband: that he would be interested in me, that he would put his store of knowledge and experience wholly at my disposal, that he couldn't be bored by the difficulties of a beginner." The writer continued, "Although it is my own letter, I think I may say that there could not be a better testimonial to your Husband's geniality and sympathy as a scientific man."[70]

Critics tended to characterize Burdon Sanderson as weak and perhaps even dim. Tyndall had described him to DuBois-Reymond as "clever and industrious, but he possesses a vague and wandering mind."[71] These opinions were most clearly revealed in debates that occurred over tactics in the anti-antivivisectionist campaign. In Foster's opinion Burdon Sanderson had "maggots in his head."[72] Much later Schäfer wrote to Burdon Sanderson at Oxford: "I am afraid that we differ fundamentally in opinion with regard to the policy to be pursued towards the anti-vivisectionists. You wish to live at peace with them; I do not: I would attack them on every possible occasion, without necessarily waiting to be attacked, and crush them."[73]

Partly because of these qualities, the Foster circle did not support Burdon Sanderson's appointment as the Waynflete Professor of Physiology at Oxford in 1883. They felt that the post needed to go to a stronger personality. Their characterization was somewhat unfair; from another perspective Burdon Sanderson could more positively be described as diplomatic. He also had to operate in very different circles than the Foster group, being at Oxford and in medical organizations in which a scorch and burn policy was not a viable alternative. Nevertheless, Burdon Sanderson's timidity was reflected in his inability to put forth bold theories.

Between Botany and Physiology: The Fate of the Venus's-Flytrap Experiments

Burdon Sanderson's 1888 Venus's-flytrap paper was the cumulation of his research, and in it he incorporated most of his previous results. It was also to be his last scientific paper on the subject. In the 1880s he developed a diphasic view of the electrical activity and thus no longer called it the negative variation. The centerpiece of the paper consisted of Burdon Sanderson's many photographs of the capillary electrometer column under different experimental arrangements. In figure 6.6 he pictured the results of four different "fundamental experiments." These photographs embody Burdon Sanderson's experimental sophistication. In order to produce them he constructed a moving photographic plate, and upon it he marked the passage of time and the point of excitation by using beams of light. The small lines occur twenty to a second; the large breaks are the point of excitation.[74]

Initially, Burdon Sanderson's investigations had been a product of their place and time. His conclusion that animal and plant excitability were fundamentally the same depended implicitly on Huxley's protoplasm theory that "the basic attributes of life are displayed by a sort of homogeneous ground substance that is common to both plants and animals and which should therefore be regarded as the most compelling evidence for the basic unity of all living forms."[75] In Huxley's usual fashion, when he articulated this theory in 1868, he was actually popularizing the work of others—notably, that of

Figure 6.6. Photographs of the fundamental experiment,
Yale University Biomedical Communications.

the German botanist Ferdinand Cohn (1828–98).[76] When Burdon Sanderson made his discovery in 1873, he was riding a crest of interest in the underlying unities between animals and plants, reflected in the large number of scientific papers published on the carnivorous plants in this era.[77]

But by the end of the 1880s these investigations went against the trend toward increasing specialization in the biological sciences. Increasingly, researchers and teachers had their own areas of expertise, and physiologists tended to ignore the plant world. This trend was reflected in textbooks. Traditionally, physiology textbooks had begun with a definition of life, which included a discussion of the differences and commonalities between animals and plants.[78] Foster opened his 1877 textbook, however, with a short discussion of life, in which he used the amoeba to characterize the organic world. In his introduction he mentioned plants only once—noting that many so-called amoebas were "in reality, merely amoebiform phases in the lives of certain animals or plants."[79] Schäfer's textbook, published at the end of the century, had no introduction at all and thus entirely omitted any discussion of life or the relationship between animals and plants.[80] Burdon Sanderson, to his chagrin (because the Oxford statutes did not allow the physiology professor to teach plant physiology), taught only animal physiology to his students at Oxford.

The new, more rigid disciplinary boundaries made it possible for botanists to reject Burdon Sanderson's data, even when physiologists accepted the data. The response of botanists to his representations also demonstrated that visual representations were not intrinsically convincing. Julius von Sachs (1832–97) wrote dismissively in his influential botany text, "I will here refer to one point only by the way." He went on to cite Burdon Sanderson's discovery—which he described as concluding "that something of the nature of animal nerves exists in the leaves of Dionæa"—as evidence of "the general ignorance which prevails as to botanical matters." "Our [botanists'] ideas of the irritability of plants, explained with so much trouble and labour, will one day be applied to the utterly obscure views as to the so-called negative variation in animal nerves," he predicted.[81]

Sachs's response demonstrated that, although these techniques had become commonplace in animal physiology, they remained foreign to botanists.[82] Botanists, whose field was fading in importance, relative to physiology, were also guarding their territory from incursions by physiologists. Sachs remarked, "In any case, then we have no necessity to refer to the physiology of nerves in order to obtain greater clearness as to the phenomena of irritability in plants." Claiming primacy for botany, he continued, "It will, perhaps, on the contrary eventually result that we shall obtain from the process of irritability in plants data for the explanation of the physiology of nerves, and this, although it is yet a very distant hope, gives a special attraction to the study of the irritable phenomena of plants."[83]

Sachs was perceptive. Animal physiologists accepted the data, but they had never been particularly interested in the Venus's-flytraps as an organism. They had hoped that the mechanism of movement in the Venus's-flytrap was analogous to, but simpler than, the mechanism in animals. Thus, they hoped to use the Venus's-flytrap as a model to learn about what had become a difficult problem. In a similar fashion DuBois-Reymond, in a paper on electrical fish, had noted that they "have indeed lost some of their wonder," now that we know all the nerves and muscles of every animal are capable of electrical actions, "but this loss is abundantly compensated by the hope which now attaches to their investigation."[84]

To retain the interest of animal physiologists Burdon Sanderson needed to demonstrate that they could learn something new about the physiology of nerves by studying the Venus's-flytrap. To interest botanists he needed to relate his observations to contemporary theories of plant movement and cellular constituents. He did neither. Therefore, his animal experiments, which were for him a sideline to his elucidation of his own discovery about plants, had a more lasting impact than did the main line of investigation. Because they were so closely related to the research of others, they were taken up in the literature and expanded upon.

From an outside perspective Burdon Sanderson's statements about the unity of the plant and animal phenomena could be seen as speculative—Sachs certainly found them to be both speculative and premature—but Burdon Sanderson believed that he was not going beyond his evidence. By *unity* he literally meant that the data looked similar, thus the representations looked similar, and therefore the underlying phenomena was somehow analogous. It was debatable how alike Burdon Sanderson's animal and plant excitability results actually were, but, more important, his conclusions lacked any explanation of this formal similarity. Burdon Sanderson's inability to relate his investigations to the larger problems in the field condemned this research to the periphery of physiology.

By contemporary standards Burdon Sanderson was successful in producing convincing representations—no trivial achievement. He noted proudly in 1899 that his results "are given in full in Biedermann's important treatise on 'Electro-physiology.'" But even Burdon Sanderson had to admit that no other plant physiologist had pursued this method of investigation, and at the turn of the century these experiments were a dead end.[85]

PART III

The Medical Sciences

Critics and Allies

Physicians, Antivivisectionists, and the Failure of the Oxford School of Physiology

Somehow or other I find that the fact that I have been thinking of the Oxford Chair of Physiology has become rather widely known and I have received a great deal of advice on the subject—all decidedly contra. The ground is chiefly that if I go to Oxford, I shall spend the remainder of my life in the struggle for the means of carrying out the organization of the teaching & leave all the fruit to my successor.

John Burdon Sanderson, 1882

In the summer of 1882 Henry Wentworth Acland (1815–1900), the Regius Professor of Medicine at Oxford, encouraged Burdon Sanderson to become the first Waynflete Professor of Physiology there. The new physiology chair was a small component of reforms at Oxford which had been undertaken since midcentury.[1] Burdon Sanderson refused to apply but agonized over whether to accept the chair if elected. The new position's attractions were: "Freedom for half the year; Means of establishing a new laboratory; Status; More leisure from having fewer lectures and fewer students; Residence out of London; Greater opportunity after the first year of doing research." On the other hand, Burdon Sanderson worried about whether Oxford would suit his health, and he rejected the status as of "no value." He would have the weight of organizing a new laboratory from the ground up, which meant that he could procure instruments, but this was offset by the likelihood that his income would diminish. Finally, he wondered whether "at 53 one ought to give up ambitious schemes."[2]

Others concurred that he was too old. In 1881 Huxley had been approached for the position, and, consistent with his opposition to Burdon

Sanderson's candidacy, he rejected it, writing, "I am getting old, and you should have a man in full vigour."[3] The *British Medical Journal* reported that "the leading physiologists of this country are unanimous in wishing the task of inaugurating a new school of physiology at Oxford to be confided to Professor Gamgee."[4] Arthur Gamgee, who was forty-one years old, was unhappy in Manchester and desperately wanted the Oxford position.[5] E. Ray Lankester, Burdon Sanderson's colleague at University College, wrote bluntly, "*You* would I think feel more keenly the annoyance of delay & opposition at Oxford & the immense dead-weight of inertia which has to be pushed against there—more than a younger man who might hope in the course of years to see his efforts bear fruit."[6] Opinion was not quite as uniform, as the *British Medical Journal* report suggested: John G. M'Kendrick, Regius Professor of Physiology at Glasgow, informed Burdon Sanderson that he would oppose Gamgee's appointment even if Burdon Sanderson refused the offer from Oxford.[7]

Burdon Sanderson sought advice from Acland and Bartholomew Price at Oxford. They explained that no expenditure on buildings or furnishings could be made without the sanction of Oxford's Hebdomadal Council and Convocation, but they were confident that there would be no problems. Acland personally pledged unequivocal support for Burdon Sanderson, stating: "You know already the dean who in such a matter is really absolute (though one must not say so in a democracy)," and "he will propose all you desire."[8] Price was equally optimistic, stating: "I in a long experience have never known a case where the University has not provided all that a Professor requires. . . . I have no reason to think that there will not be provided for the new Physiological Professor such accommodations as he requires and on his own plans."[9]

Despite Burdon Sanderson's coy refusal to apply for the position, he was clearly the inside candidate—a fact that was resented by Gamgee's supporters. The election was delayed several times due to the absence of Moseley, one of the electors and a supporter of Gamgee, in order to preserve appearances. Thus, it was no surprise when Burdon Sanderson heard on 23 November 1882 that he had been elected the first Waynflete Professor of Physiology at Oxford.[10] Acland, who had arranged Burdon Sanderson's appointment with the aim of rivaling Foster's school of physiology at Cambridge, was pleased. He wrote, "My desire for nearly 40 years has been to see here laid the Scientific basis of Medicine—through Physics, Chemistry, & Physiology each in their widest sense but in correlation."[11] Huxley, voicing the feelings of those who had cautioned Burdon Sanderson about Oxford, wrote less effusively that he did not "know whether to be glad or sorry and you seem to be in like case."[12]

Huxley and Lankester were prescient. Despite the efforts of Burdon Sanderson and Acland (among others), the new physiology program and the related revival of preclinical medical teaching were both failures at Oxford, at least in student numbers, the main criterion by which it was judged at the time. The failure of the Oxford school of physiology in the 1880s and 1890s was due less to the opposition of clerical and classical obscurantists than to a combination of the prominent British antivivisectionist movement and the much smaller circle of Oxford-educated medical practitioners. Contending interests within the scientific professorate, described further in the next chapter, also contributed to its failure. While the opposition of antivivisectionist to physiology was obvious, events at Oxford demonstrated the lack of support among medical practitioners for a move to a science-based medical curriculum in late Victorian Britain—an attitude found among the British medical profession generally. In fact, the actions of the Oxford medical graduates and the wide sympathy they found in the medical community illustrated the growing opposition to the ideals of science among elite practitioners. This group of medics, in reaction to the ascendance of laboratory research as a defining activity for the medical profession, rejected the ideals of science, in favor of older gentlemanly ideals.

Who were the Oxford medical graduates? They tended to be London consultants. Doctors had been for some time divided between general practitioners and the elite hospital consultants. The most common route to general practice at midcentury was still to become a licentiate of the Society of Apothecaries and a member of the College of Surgeons. Consultants tended to be academically trained, to be members of the Royal Colleges, and to practice in or close to London.[13] An Oxford medical degree was a route to such a career.

Burdon Sanderson's career at Oxford was carried out in the shadow of Michael Foster's School of Physiology, as it was called by contemporaries. The success of a research school depends on three factors: the personality of the leader; the research program itself; and support, both institutional and financial.[14] The last chapter outlined Burdon Sanderson's failings with respect to the first two factors; this chapter focuses on the institutional and financial support, or rather its lack, for physiology at Oxford in the 1880s. Widely recognized at the time, numerous contemporary writers refer to the failure of Oxford physiology, particularly as compared to Cambridge. But the "failure" was a relative one; had it been compared to provincial programs, some of which had appointed physiology professors in the early 1870s, Oxford would have been a success. But for contemporaries the only comparison made was with Cambridge—while Burdon Sanderson himself missed the members of the flourishing department he had left behind at University College London.

In the 1880s all British physiologists, like Burdon Sanderson, struggled to establish their discipline as an experimental, medical science. In the British context the 1870s were a turning point in the institutionalization of physiology. Notably, Owens College, Manchester appointed the Brackenbury Professor in Physiology in 1873 (Arthur Gamgee); Edinburgh University appointed a professor of physiology in 1874 (William Rutherford); University College London appointed the Jodrell Professor of Physiology in 1874 (Burdon Sanderson). Michael Foster had gone to Cambridge in 1870 as the Praelector in physiology at Trinity College. And, before the creation of the Waynflete professorship in 1883, King's College, London (Gerald Yeo, professor, 1880), and the Birmingham Medical School (Alfred Carter, lecturer, 1881) had created physiology positions.[15]

The relationship between physiology and medicine was complex. It could be viewed, the historian Gerald Geison has proposed, as symbiotic, with physiologists depending on medical schools to provide them with both an audience and necessary resources and physicians in turn depending on physiology for useful theories, techniques, and instruments. Or the physiologists could be seen as parasites, profiting from their association with medical education without offering anything substantive in return. Medical historian S.E.D. Shortt has suggested that perhaps the symbiosis was subtle—even if physiology were not directly useful to medical practice, the place of physiology in the medical curriculum allowed physicians to use the rhetoric of science to bolster their own professional aims.[16]

More recently, historians have argued that these explanations assume that a monolithic medical profession had embraced science, whether it be for rhetorical or practical reasons. These historians argue that a significant number of practitioners resisted the ideals of science: some were ambivalent, and others were openly hostile to the claims made for the relevance of physiology to bedside practice. The skeptics looked to alternative sources of clinical authority: empirical observation (rather than experiment) and character. Some cultivated gentlemanly ideals.[17]

The failure of the Oxford physiology program was due in part to the support of gentlemanly over scientific ideals. Gentlemanly ideals were rooted in the complex British class system; specifically, they embodied the hopes that some physicians could be considered members of the upper class. To be a gentleman was to be a man of good breeding and good social position—that is, a member of the upper echelon of British society. Since even elite physicians were not men of leisure (one criterion for gentlemanly status), they had to demonstrate their fine feelings and their generally elevated position in other ways. A gentleman was also cultured, signified by his attainment of a classical, liberal education. In contrast, those who espoused scientific ideals were drawing on the ascendant middle-class professionalism, which rooted its claims for status in expert knowledge that could be systematically deployed.

The elite London practitioners who were the Oxford medical graduates did not feel that their professional status rested on their claims to the mastery of natural sciences or that their status was best affirmed by deploying the rhetoric of science. As they saw it, the basis of their elevated status was their classical education, which marked them as cultured gentlemen.[18] Not coincidentally, the Oxford medical graduates were aware that their education had required significant time and finances; thus, their attainments were—and should remain, they felt—the exclusive preserve of that small fraction of the medical profession with sufficient means and leisure. Other, more disinterested observers also deplored the move at Oxford to allow aspiring physicians to choose a scientific curriculum rather than the traditional arts curriculum for their first degree. The upper ranks of the medical profession had misgivings about the new trends in medical education. Many hoped that a small number of physicians could continue to display the credentials of gentlemen and thus uplift the entire profession in the eyes of the public.

In many ways this conflict was about character. There were two related issues: What methods properly molded medical students into practitioners? And what credentials signified to outsiders the integrity of a practitioner? A classical liberal education had traditionally done both for the upper ranks of the medical profession. Those who espoused the ideals of science were challenging the older gentlemanly ideals and supporting a meritocracy. Thomas Bonner wrote of the mid-nineteenth century, "Leading physicians in Britain and America continued to insist on the importance of a classical education, good manners, and the ability to relate to the privileged classes for those who sought eminence in their work."[19] The content of the debate also underscored the masculinity of the Victorian medical profession: both gentlemanly ideals and the ideals of science could be used to exclude women from medical practice. Women were not eligible for the upper ranks of the profession, although by the 1890s there was a movement supported by some leading practitioners, including Acland, to train women to go out into the Empire as medical missionaries.[20]

The medical profession publicly did close ranks to oppose the antivivisectionist movement, which sought to eradicate animal experimentation; nonetheless, there was an intersection between the beliefs of those who relied on gentlemanly ideals and those opposed to animal experimentation. Antivivisectors believed that vivisection as an activity was morally wrong and corrupted those who engaged in it. In particular they argued, and on this point they garnered some agreement from the members of the medical profession, that vivisection as a part of medical education was wholly inappropriate because inevitably it produced practitioners who were hardened to suffering and pain.[21]

The failure of the Oxford school of physiology was also due to ambiguities underlying the campaign for a scientific medicine. Not all practitioners agreed with Burdon Sanderson that physiology was the most important of the medical sciences. The term *scientific medicine* was so broadly defined that supporters often quarreled over specific endeavors related to it. The diverse activities that Burdon Sanderson himself had carried out under the rubric of science demonstrated the slipperiness of a medicine based in science. At Oxford, Acland considered pathology to be more fundamental to medical practice than physiology, and he did not entirely support animal experimentation. Others considered clinical training to be the formative part of the medical curriculum.

Confusion about what a department of physiology should be further contributed to the failure of the Oxford department. A physiology department could merely offer lectures and perhaps a practical course, or it could constitute a research school—that is, a department that had as its central mission original investigations designed to make discoveries that would, as Burdon Sanderson put it, "lay a brick which will remain part of the building [of knowledge]."[22] Thus, at Oxford medical graduates could sincerely claim to support scientific medicine in general and physiology in particular while making it impossible for Burdon Sanderson to muster the resources necessary for physiological research—resources that would come only with large numbers of students. In addition, within Oxford there were long-standing and unresolved debates about the role that research could or should play in the university.

Oxford and the Natural Sciences

Upon his arrival at Oxford in 1883, Burdon Sanderson enthusiastically launched a three-part plan for reform: beginning a process that involved a myriad of committees, memoranda, and meetings, he strove to set up the physiology program (both the teaching and the physical plant), change the examination structure, and revive medical teaching. This three-point program was necessary because Burdon Sanderson realized that merely founding a physiology department was not likely to be successful. It was well known that Oxford's examination structure put it at a disadvantage relative to Cambridge in terms of attracting science and mathematics students. Both Oxford and Cambridge had many quaint and, to the uninitiated, confusing titles for their various examinations. At Oxford the matriculation examination that a student was expected (in theory) to pass upon coming up—that is, on arrival—was "Responsions"; a year after passing Responsions the student needed to pass his "Moderations" examination. Science students then had to pass a general examination in the natural sciences, "Prelims," and, after passing a final examination in the Natural Sciences School, they would be awarded a degree. At Cambridge the matriculation examination was the Previous, and since 1881 the second examination for a science student was the first half of the Natural Sciences Tripos; after passing the second half, the student would be awarded a degree.[23]

By the 1880s the proportion of Oxford graduates pursuing science degrees—that is, "reading the Natural Sciences School"—had fallen dramatically relative to the numbers "taking Cambridge's Natural Science Tripos." Historian Janet Howarth compared the numbers of the two universities:

Cambridge

Old examinations	1851–81	405
Part 1 only	1881–1904	1,598
Part 2 only	1881–1904	490

Oxford

	1855–79	383
	1880–1904	903

Women were allowed to take examinations but could not receive degrees. At Cambridge, between 1881 and 1916, 299 women took part 1 only, and 104 took part 2 only. At Oxford one woman took a natural science examination.[24]

The largest barrier to potential science students at Oxford was the requirement that they pass two classical examinations before being eligible for an Honours degree program, such as in Natural Sciences. Oxford's matriculation examination was reputed to have a higher standard for Greek than that of Cambridge. Students from outside the public school system—that is, those who had not been classically educated—often had to study a further year at Oxford in order to pass this matriculation exam. They were then required to wait a year before taking Moderations, another classical examination. At Cambridge, in contrast, having passed the matriculation examination, any science student had fulfilled the classical requirements of the degree and could begin preparing for the Natural Science Tripos.[25]

In 1885 Gilbert C. Bourne, who would be appointed the Linacre Professor of Comparative Anatomy in 1906, wrote in the *Oxford Magazine* that "very few men can afford to throw away two years at such an expensive place as Oxford; and if they can afford the money they cannot afford the time." Bourne identified the Moderations examination as the crux of the problem. He supported and recommended substantive changes to this particular exam so that the plant, faculty, and expenditures of the biology departments would no longer be wasted on "a few enthusiasts" and "Cambridge with its 250 biological students and flourishing medical school would cease to be a reproach to us."[26]

Huxley also supported such changes, writing to Jowett to suggest a letter to the vice chancellor in which "we prudently ask for the substitution for modern languages (especially Greek) and elementary science for some of the subjects at present required in the literary part of the examinations of the scientific and medical faculties."[27]

Because many science students in late Victorian England intended to be physicians,[28] the minimal opportunities for preclinical or clinical medical studies at Oxford further discouraged their attendance. The university did grant medical degrees, but those degrees were for the most part by examination only. In order to be eligible for an examination the practitioner had to be the holder of an Oxford B.A. degree. After receiving the degree, most likely in a subject wholly unrelated to the practice of medicine, the student would go elsewhere for medical training, usually to one of the London hospital schools; the small number of men who received Oxford medical degrees were those who then returned to Oxford merely to sit the required examinations. Between 1861 and 1870 there were twenty-eight Oxford M.B. degrees awarded and forty—nine at Cambridge; between 1871 and 1880 there were forty-five at Oxford and eighty-three at Cambridge.[29]

At Cambridge, in contrast, many science students were able to fulfill medical licensing requirements as part of their B.A. degree program. Speaking as a naturalist and a father of five sons who needed to support themselves, William Benjamin Carpenter (1813–85), the author of *The Principles of General and Comparative Physiology,* summed up the problems in a strongly worded letter to Acland: Oxford lacked a "suitable *entrance* Examination"; the university needed to design a "three years' Course of Physics, Chemistry, and Biology" which would also "constitute a first-rate preparation for Medical Study."[30] If this course were followed, Carpenter stated, "I have no doubt whatever that [Oxford] will attract a high class of Students, who would do credit to the University, and become as conspicuous in the Scientific world as the Cambridge Science men are now." He emphasized that, despite all that Oxford had to offer, as he had done with his son Philip Herbert (1852–91), "I should feel it necessary, if I had another son . . . to send him by preference to Cambridge; not being able to afford to let him spend two preliminary years in pursuits which would be of little or no after service to him."

Burdon Sanderson's plans depended on the support of the various governing structures, but he soon discovered that Acland and Price had misled him about the power structure of the university. Oxford was governed by a combination of oligarchy and democracy, and power was not centralized. Each college was an autonomous institution, with its own revenues, run by its own fellows, who elected any new fellows and the head of the college. The heads of the colleges could have various titles: principal, warden, master, rector, president, or provost. The head of Christ Church was also dean of the Cathedral, and thus Liddell, its head, was known as Dean Liddell. The college heads served in rotation as vice chancellor of the university.[31]

The university as a whole had three governing bodies. At the top of the hierarchy was the Hebdomadal Council, which consisted of twenty-two members: the chancellor, the vice chancellor, the two proctors, and eighteen elected members. The elected members were chosen by a ballot of Congregation, which consisted of all professors, some university officials, and all members of Convocation who lived in Oxford. Oxford graduates were members of Convocation and, if they chose, could attend any meeting and vote.[32] The Hebdomadal Council wrote and circulated university legislation—the ubiquitous statutes. Any statute was then debated in Congregation, which could accept or reject it. The statute, if accepted, then proceeded to Convocation, which could also only accept or reject it.[33]

Men like Henry George Liddell, dean of Christ Church (the dominant Oxford college and Acland's college) and Bartholomew Price usually controlled the Hebdomadal Council. Liddell and other influential Oxford figures, including H. S. Smith, Price, and Benjamin Jowett, had strongly backed Acland's earlier attempts to reform science education, which fit with their own vision of a reformed university, providing training for the professions. Liddell, who was increasingly impatient with Acland's slow progress in reviving medical teaching, was a strong supporter of the new physiology professor.[34] This group's long-standing irritation with Acland was reflected in Benjamin Jowett's blunt remarks in an 1869 letter to Florence Nightingale: "Dr. Acland is called 'Barnum' at Oxford. He is one of the vainest, rudest men that ever lived. He is also one of the greatest bores that ever lived. But against his boring must be set that he has got one or two good things done; & against his rudeness that he is extremely kind to some persons. I believe him to be neither a man of science nor a good practitioner. He is an intolerable ass & an unendurable bore, yet a worthy man after a fashion. What could make you waste 8 hours upon him? This shows me how you are at the call of everyone."[35]

The clerical element, which had once dominated Congregation, was no longer so prevalent. Nor was the center of opposition to the new physiology program clerical. In 1894 Burdon Sanderson wrote to Edward A. Schäfer, his old colleague from University College, now also a prominent British physiologist, to correct his misconception (a common one) that clerical opposition had been important in impeding the sciences at Oxford in the 1880s: "As regards the word 'clerical' the leaders of the high church parties here have not been opponents—Paget is a reliable supporter and so was Talbot. The former Dean of Christ Church was ever a strenuous supporter & made no reserves. On the other hand our most 'virulent' enemies have not been clerical."[36] And, in fact, Congregation never created any difficulties for Burdon Sanderson.

Burdon Sanderson discovered the democratic nature of Oxford, however, when the funding for his department—initially viewed as the least controversial part of his plan—was threatened by antivivisectionists, who tried, in Convocation, to block the financing for the new physiology building. E. B. Nicholson, the Bodleian librarian, and E. A. Freeman, fellow of Oriel College (and from 1884 the Regius Professor of Modern History), led the antivivisectionist party at Oxford. This group was, of course, part of a larger British movement, and, in fact, the majority of the Oxford antivivisectionists seemed to be no longer active members of the university.[37]

The British Antivivisectionist Movement

In the early 1870s the British public had become aware that its scientists, like those on the Continent, regularly performed experiments on live animals. Various organizations campaigned against these "vivisectors," as they called them. The publication of the *Handbook for the Physiological Laboratory* in 1873, which Burdon Sanderson edited, became a focus for their anger. Under the de facto leadership of Frances Cobbe the British antivivisection movement emerged as a potent political force. In 1876 her group, the Society for the Protection of Animals Liable to Vivisection, spearheaded the passage of an act to ban vivisection which was not entirely successful. After intense lobbying by the scientific community, the 1876 act merely regulated the practice. It required researchers to obtain licenses to perform vivisection experiments, banned the use of such experiments for educational purposes, and restricted experiments performed without analgesics. Nonetheless, the highly publicized antivivisection campaign, which initially took the proponents of physiology by surprise, was painful for Burdon Sanderson and other researchers who were personally attacked.

It was no coincidence that the public face of the antivivisection movement was female, for Victorians women were the repositories of the humane, morally uplifting values of the community. The gender dynamics of antivivisectionist sentiment were also subtly revealed by Ghetal. She assisted John frequently and even helped him set up for experiments, but she did not participate in animal experiments. She revealed on many occasions that she fully understood in general terms what it was that John did in the laboratory, and approved, but she chose not to dwell on the details. She too suffered some public opprobrium because she was John's wife: in 1881 Ghetal's appointment to the council of Somerville Hall, a women's college at Oxford, was opposed by Cobbe, despite Ghetal's long-standing interest in the education of women.[38]

Over the years Burdon Sanderson had experimented on countless animals: guinea pigs, dogs, frogs, and cows. On 5 August 1870 he noted in his diary that he was engaged in experiments with Brunton on the movements of the heart in the recently born puppy. The matter-of-fact way Burdon Sanderson recorded these experiments, as well as the much-criticized tone of the *Handbook,* demonstrated how out of touch he was with general British sentiment toward animals. He was both shocked and hurt by the personal attacks of the mid-1870s and from that point on was hesitant to flout public opinion openly. After his public lecture at the Royal Institution in 1881, Ghetal wrote in her diary: "Mr. T. was so much amused at J. calling his preparation 'a cone of protoplasm' when it was really a frog's heart. He had, Mr. T said, evidently the fear of Miss Cobbe before his eyes."[39] Increasingly, in his own research Burdon Sanderson focused on the electrical activity of the Venus's-flytrap, and, although a wit did suggest that the antivivisectionists might take the plant up, this research choice was clearly much less controversial.[40] We can only imagine Burdon Sanderson's horror when the antivivisection movement emerged publicly again to attempt to block the founding of the Oxford physiology department.

While the supporters of physiology argued that vivisection contributed to knowledge and led to medical innovations that made humans healthier, the desire to accumulate knowledge did not resonate with the public. Members of the antivivisection movement, while attempting to prove that animal experiments had not assisted therapeutics, also entirely rejected the move toward a medicine based in the research sciences on the grounds that it would be immoral. They were not, by and large, supported by the medical profession, although there were some practitioners who sympathized and a few who actively joined the campaign. The medical sympathizers agreed that the humanitarian values of medicine were of primary importance for practitioners; some also rejected physiology as irrelevant to practice, championing clinical experimentation as the important science for practitioners.[41] Thus, although few in the medical community could embrace the radical rejection of medical science of the antivivisection movement, they did not necessarily concur with Burdon Sanderson's vision of the medical sciences, in which physiology played a pivotal role. The antivivisectionists both made such schisms in the medical community evident and exploited them for their own purposes.

The Establishment of the Physiology Department

Convocation in June 1883 voted to fund the new physiology building, but with a majority of only two votes. The antivivisectionists tried again in February 1884 and March 1885 to deny necessary funds to the new physiology department. They were defeated, the physiology building constructed, and the department funded, but the votes in Convocation were far from formalities. Moreover, most votes were cast in favor of the building in order to show opposition to interference by outsiders in Oxford affairs, rather than to support physiology. As Burdon Sanderson put it, "There are many people who do not like experiments on animals who are nevertheless glad that any attempt to interfere with the internal management of the University should be defeated."[42]

The fight to abolish Moderations, the second classical examination, was equally contentious. In 1885 the Hebdomadal Council began considering a proposal to replace Moderations with preliminary examinations for all of the Final Honour Schools such as physiology. Everyone recognized that the Oxford curriculum was problematic, particularly for science students, but many nonetheless opposed the abolition of Moderations. The intended changes meant that the university was reducing its commitment to the classical liberal education, although it was not eliminating the classics. The proponents of the new scheme were themselves somewhat ambivalent and were careful to emphasize that all Oxford graduates would continue to demonstrate a knowledge of "general culture."[43] This ambivalence was reflected by the editors of the *Oxford Magazine,* who supported the reforms but with reservations. Acland also opposed the plan: as Burdon Sanderson recounted, "Acland has declared himself an opponent and came to tell Ghetal that he was so and would win."[44]

Acland had been the dominant force behind the development of science education at Oxford, but in many ways he was a traditionalist. He opposed any attempts to narrow the curriculum but supported its expansion by the addition of natural sciences. In particular he believed that physicians should take their first degree in a literary, historical, or philosophical field. He rejected the general sentiment that students could no longer become proficient in the ever-expanding sciences in the year or two that remained in their curriculum after the mandatory classical studies.

In this atmosphere the reformers became worried when the preamble to a statute proposing to abolish Moderations was defeated. The debate over this preamble revealed how the reformers and antireformers of Oxford were polarized on a variety of related issues. The participants aligned themselves as in the antivivisection debate: Price introduced the measure, and it was roundly opposed by Freeman. Freeman was not only an antivivisectionist but also a long-standing opponent of modern reforms at the university. He stated that it was the role of the university to give a general education, not to educate for the professions (this was a clear attack on the liberal campaign to create a modern role for the universities as the educators of the professions). The reformers had the last word: in the fall of 1885 Moderations were abolished for students who planned to enter one of the Honours Schools.[45] Nonetheless, it was clear that Oxford, unlike Cambridge, was unwilling to leave education in general culture to the schools.

This reform victory did not mean that Oxford had wholeheartedly embraced the science programs; the opposition remained and perhaps even grew as the costs associated with science education were realized. The *Oxford*

Magazine reflected the ambivalence of the university. Although it had strongly supported the new physiology building (and the expansion of other science departments), it editorialized in 1885: "Here in Oxford we must always remember that endowment for research, pure and simple, has been tried and condemned. . . . There are many other ways of aiding a science or justifying an endowment than the production of monographs on points 'introduced by Germans.' The best researches . . . have been those of men who had . . . something beside the abstract self-dictated investigations of their laboratory or study."[46] In 1886 the magazine noted that the "the annual cost of educating an undergraduate in . . . [the] branches of science is enormous" and implied that the university should concentrate on the less expensive "other studies for which Oxford is more famous."[47]

A letter to E. B. Poulton (who became the Hope Professor of Zoology in 1893), which he passed on to Burdon Sanderson, indicated that such sentiments were not unique to the *Oxford Magazine*. Paul Willert, a history examiner (and thus a likely supporter of modern research), nevertheless wrote, "When a friend takes the trouble to write a letter it is hard to say no—but honestly there are 50 objects to which I would sooner give my money than to the dissecting shed, & to wh[ich] in all possibility I shall *not* give it!" He continued, "Having sucked alma mater dry . . . scientific people are now trying to drain our private pockets!—& I fear all to no purpose so far as the medical school gives till you can dissect Acland."[48]

Oxford remained hostile to science—although, as Janet Howarth has pointed out, this hostility has been exaggerated. Many of the problems the science departments experienced were due to institutional arrangements that were unfortunate but were certainly not planned to stymie scientific research. One element was the settlement of the mid-Victorian Parliamentary Commissions in the statutes of 1882, which left the university at Cambridge with a larger income and a free hand to spend it without reference to the colleges. Ironically, this situation had come about because at the time of the commissions it was the poverty of Cambridge which was most prominent. This important difference allowed Cambridge to create positions in the sciences, while at Oxford this responsibility was left to the colleges, which usually abdicated it. Thus, by 1886 Cambridge had eighteen science professors overall to Oxford's thirteen but, more important, three times as many lecturers and readers.[49]

The depth of feelings against science studies, and in favor of the tradi-tional curriculum, was reflected in the difficulty experienced in amending those institutional arrangements that hampered the science departments. Despite an early start, by the 1880s there was a significant party within Oxford which was balking at the costs associated with science instruction and was willing to allow Cambridge to overtake Oxford as the university of choice for students in the sciences.

The difficulties that the science departments faced also reflected the splin-tering of the Oxford reform movement.[50] In the 1850s groups with diverse aims had been able to coalesce around a common goal—the need for reform. The end (in 1871) of the Tests, which had required degree candidates to swear to orthodox Anglican tenets offensive to many Dissenters, left liberals less interested in further Oxford reform.

The continuing agricultural depression (which had begun in the late 1870s) imposed financial constraints on the university which damaged the coalition between the scientists and those in the humanities who supported the research ideal.[51] Their divergence was illustrated by Willert's colorful ob-jections to the "dissecting shed." Willert's antiscience sentiments reflected the conclusions of those in the humanities that, with limited funds available, science could thrive only at the expense of their own fields of interest.

The inner circle that ruled Oxford—men like Jowett and Liddell—con-tinued to support science; for them it was necessary for Oxford's reputation and perhaps even the good of the nation that science flourish in the univer-sity. And the changes to the examination structure and the funding of the physiology program did occur. The antivivisectionists and traditionalists were unable to block Burdon Sanderson. But it was this other opposition to the physiology department which in the end allowed the medical graduates of Oxford to obstruct the development of physiology, by blocking the creation of a vibrant medical school at the university.

Controversy over the Creation of the Oxford Medical School

In the fall of 1884 Burdon Sanderson began to collect information about Cambridge's medical program and the requirements of the Royal Colleges of Physicians and Surgeons.[52] The two Royal Colleges had collaborated to form a single examining board to administer the new "conjoint examination" which was to take effect in 1886. Burdon Sanderson took this course because it was common practice for Oxford B.A. degree holders who planned to take the Oxford medical degree (the M.B.) first to fulfill the requirements for membership in one of the Royal Colleges—in order to be able to register and thus practice before they received their degree. The holder of an Oxford medical degree was, of course, permitted to appear on the registry of medical practitioners and thus could legally practice medicine. As was already mentioned, medical training at Oxford was wholly separate from the undergraduate degree program, although keen undergraduates might have received some medical education at particular colleges and at the Radcliffe infirmary.[53]

When Burdon Sanderson inquired about the status of Cambridge as a recognized medical school, Donald MacAlister of Cambridge explained the consequences: the Royal Colleges accepted their lecture courses for admission to examinations, and the period of enrollment at the university counted toward the required period of study. Their students also enjoyed a series of exemptions from examination; the second Cambridge M.B. examination, for example, was accepted in place of the primary examination for membership in the Royal College of Physicians.[54] Sir Henry Pitman, secretary of the Royal College of Physicians, informed Burdon Sanderson that, since Cambridge and Oxford followed the same academic calendar, if Oxford were to institute a similar curriculum, it too would likely be designated a "recognized" medical school.[55]

A student entering Oxford in 1884 required at least eight years to obtain a medical degree. In contrast, a Cambridge student could pursue his medical studies without a B.A. degree, potentially reducing the length of the program to the legal minimum of five years.[56] At the beginning of 1885 Burdon Sanderson planned radical changes to shorten the time required to obtain medical degrees at Oxford. He hoped to have the Oxford science program accepted by the Royal Colleges and in exchange proposed that candidates who had passed the new conjoint examination of the Royal Colleges be exempt from parts of the M.B. degree examination process.

Oxford medical graduates became aware that changes were being contemplated. Acland apparently revealed the nature of the proposal in statements at the Royal College of Physicians.[57] Almost immediately a letter appeared in the *Pall Mall Gazette*, "A Word of Warning to Oxford Men," by an anonymous B.A. who was against the plan.[58] Opposition was spearheaded by the Oxford medical graduates' club, which had begun the year before as a social organization.[59] The club had been organized mainly by elite practitioners who also taught in the London schools. In 1884 the *Lancet* had recorded the inaugural meeting of the club attended by Burdon Sanderson, Acland, and about fifty-four others. The note stated, "The object being entirely social, the speeches were limited in number."[60]

The graduates objected: it was inconceivable that Oxford would accept the conjoint examination as a part of Oxford's degree requirements. After all, the conjoint examination was to be "the *lowest standard* with which it is considered safe to let General Practitioners loose," according to Frances Champneys, an Oxford M.B. and examiner for the conjoint examination.[61] J. Matthew Duncan concurred: "A University degree should easily be in all respects . . . higher than a corporation license. Or rather it should be much higher in all but practical subjects."[62] Their strong objections were not unusual. Burdon Sanderson's proposal was revolutionary. It was not the practice at Cambridge or even the University of London to accept the colleges' examinations. G. Humphry, who held the surgery chair at Cambridge, was dubious about the scheme, writing to Acland that he was surprised, "at Sanderson's suggestion & that I hope it will not be acted for that a Univy [*sic*] degree should imply something more than a qualification to practice."[63]

In February 1885, as another Convocation vote for funds for the physiological laboratory was approaching (on 10 March), Burdon Sanderson began receiving letters from Oxford medical graduates. These letters contained threats that, if the proposed reforms of the medical school were unsatisfactory, the graduates would no longer support the vote. Champneys wrote, "If the standard of the conjoint Board (which it must be remembered is the lowest standard consistent with a qualification to practice at all) is accepted in whole or even in part for the final M.B. of Oxford, a very serious blow will have been struck at the prestige attaching to the degree." He then warned Burdon Sanderson, "In such a case I fear that the interest which all now take in the question of medical education at Oxford, including the study of Physiology, would be greatly diminished in the majority, if not actually extinguished in not a few."[64] Another clinician and Oxford graduate, Samuel West, stated, "You know how loyal the Oxford graduates have been to you & Sir Henry &

I do hope that they will be taken into the confidence of the authorities." He continued, "It is also quite clear to me that mischief may come if the present misapprehensions are allowed to continue uncorrected and may even from what I have heard influence the vote on the 10th of March though I hope this will not be the case."[65]

These letters also reveal more self-interested motives. As West put it in a letter to Acland, the graduates were afraid of "Cambridgizing our M.B. and so destroying its prestige."[66] What did "Cambridgizing" the degree mean? The *British Medical Journal* stated, "The experience of Cambridge has taught us that there is a great liability that young students may be induced, by the injudicious zeal of teachers of the preliminary sciences, to devote too much time to subjects which have, at least in their more elaborate details, only a remote bearing on the medical art."[67] The journal deplored the tendency at Cambridge to study the preliminary sciences for their own sake—a contention with which the Oxford graduates would certainly have agreed: as they saw it, the prestige of the degree would be destroyed by shifting time to the preliminary science subjects and away from literary subjects and professional training. Yet one cannot escape the suspicion that the prestige would be even more cheapened by making the degree common—that is, by increasing the numbers of men who held an Oxford medical degree, thus diluting its value.

The aims of the graduates were diametrically opposed to the needs of the physiology professor, who was unlikely to attract sufficient numbers of students unless he could promise that they could simultaneously complete a full preclinical medical education. In the 1850s Acland had managed to persuade the university to make large expenditures for the sciences without large student numbers, but that era had ended. In the austere 1880s "popularity," as demonstrated by student numbers (and fees), was required for new expenditures. The opposition was made clear by Champneys, who stated that it was even more important to maintain the general cultural standard than the professional standard (also endangered by the proposal), and "to maintain it, time is required . . . and short cuts cannot be allowed."[68] Yet students were not going to choose Oxford over Cambridge for medical (or physiological) studies unless the time it took to earn a degree and professional qualification was significantly shortened. Ironically, it appeared that the main supporters of the physiology program were going to doom it. Burdon Sanderson quieted the graduates by explaining that the proposal was not yet formulated and the matter would not be dealt with until the fall at the earliest, and he won the March vote on funding for physiology.[69] But Acland had been shaken by the rancor of the debate and the opposition of old friends.[70]

Burdon Sanderson still hoped to succeed in his reforms of the medical statutes. On 9 June the preambles of two statutes he had spearheaded passed Convocation. The first created a medical faculty (previously, any medical professors, like Acland, had been part of the Natural Sciences Faculty Board), which was uncontroversial. The second was Burdon Sanderson's radical proposal that Oxford recognize in part the examinations of the Royal Colleges; this change to the medical statutes was "unanimously opposed" by the Oxford medical graduates.[71]

The Oxford medical graduates were in the rather delicate position of opposing the use of examinations to which they sent their own students and over which they presided. Their published remarks were carefully worded so as to imply not that the college examinations were unacceptable but merely that they were not of an Oxford standard. The graduates used the traditional techniques of protest within Oxford: petitions, printed memoranda, and letters, which they circulated within Convocation.[72]

Liddell wrote to Burdon Sanderson on 12 June indicating that the medical graduates were continuing to be a problem: "I read your letter to the Council today. But the Amendments handed in the other day were so numerous & so complicated, that the Vice Chancellor has not had time to consider them, in order to report to Council (as the Statutes require) which were in order & which not." Liddell attempted to see the positive side: "It is not likely that the Vice-Chancellor can get the Medical Committee of the University to sit again till after Long Vacation. Perhaps it will be as well. The graduates may cool down a little & think, when they see the Statute a little more closely, that there is less to object to then they imagine."[73]

The graduates did not cool down. By this point Burdon Sanderson's enthusiasm was gone. He wrote at the beginning of 1886, "I am beginning to be very tired of the whole business and wonder whether if the school flourishes and the number of students is doubled, I shall rejoice in our success or the contrary."[74] Faced with the opposition of the medical graduates and the support they had mustered, he abandoned his proposal for sweeping changes. He was clearly mindful that he could not afford to offend the core of his support in Convocation as he faced ongoing opposition from the antivivisectionists (and traditionalists).

As late as 1891, Burdon Sanderson, while recovering from an attack of flu, worried that the funding of the new anatomy department would be threatened by the traditionalist-antivivisectionist coalition. At that time he wrote to Acland:

> I have endeavored to learn what my former opponents intend to do, but no one seems to know.
>
> I suppose that Prof. Freeman will make a speech which I hope will be violent & personal. And somebody else may ask some foolish questions.
>
> Unless some misstatement of fact is made, neither speech nor questions need be answered and I should for my own sake be glad that nothing should be said about experiments.[75]

These fears, however, appear to have been unwarranted.

Returning to the 1880s, in their final version the new statutes, which were to take effect in 1887,[76] were substantially watered down to satisfy the graduates. The new faculty of medicine would have no reciprocity with the conjoint board. The degree requirements were slightly altered so that a student could sit the first M.B. examination one year after obtaining his B.A. degree and the second M.B. examination two years later. With the changes in the physiology program and the abolition of Moderations, it was more likely that a student could receive his degree within seven years (women were still not eligible for degrees). Nonetheless, the program remained substantially longer than that of Cambridge—not a trivial difference in a period when a student's guide to Oxford devoted an entire chapter to the "Expenses of Oxford Life," with a featured section entitled, "An Oxford Career Not Cheap."[77]

The debate over the new medical school revealed the ambivalence that existed about the physiology program and the relevance of science to medicine. The medical graduates' main reason for supporting the program had been the sense that Oxford was falling behind Cambridge, alongside the belief that physiology was a necessary part of medical training and probably a good thing. But they did not really understand the extent of financial resources which a research school in physiology would require, nor did they directly connect physiology to medicine. Their support ended with any change that might affect their elevated status.

Champneys voiced most clearly the beliefs of Oxford medical men that their most important attainment was general culture (achieved at school and Oxford); second came professional qualifications (achieved through their clinical training); and last was scientific knowledge (achieved through pre-clinical studies now possible at Oxford). For them physiology, which had the pure aim of discovering "the valuable knowledge . . . which is permanent," was merely a secondary method for building character.[78] In their opinion the most important visible sign of character for Oxford medical graduates re-mained, as Champneys stated "their *general* attainments and sound and wide education, which not only marked them out but enabled them to deal with technical problems in a manner unobtainable by those whose education had been less liberal and extended."[79]

Burdon Sanderson, who continued to attend musical concerts when pos-sible, was not hostile to culture as such and believed that studying physiology built character. But he was in a minority in believing that physiology had the potential to alter medical practice. The potential utility of physiology played little part in these discussions; in principle Oxford graduates had no need to concern themselves with such matters. In reality the next generation of doctors were quite willing to leave the attainment of general culture to their school years; at university they wanted to acquire the required scientific back-ground while obtaining their professional credentials as quickly as possible — they needed to support themselves. The result was that the numbers of stu-dents in the biology departments remained small.

The Oxford medical graduates were not ordinary practitioners, and their aims were not those of the profession as a whole. Their campaign can be seen as a reaction to attempts to use physiology as a new, more democratic basis of status for medical practitioners. After all, anyone could learn physiology — given the proper lectures and training (which were becoming increasingly available) — but only a few should hope for an Oxford medical degree, with its stamp of being a gentleman.

CHAPTER EIGHT

A Corner Turned?

Experimental Medicine in
Late Victorian Britain

... a beginning of a last chapter.
Burdon Sanderson to F. J. M. Page, 1895

John Burdon Sanderson believed that experimental physiology and pathology would change medicine for the better. To practice medicine effectively, he contended, practitioners needed not only to grasp these sciences but also to understand the basics of chemistry and physics. Burdon Sanderson's vision of scientific medicine was not unique. His two most important mentors, Simon and Acland, shared many of his views. Believing that physiology was essential to pathology, Simon had supported the reform of medical education and, as we have seen, strongly supported pathological research in Britain. Acland had campaigned for the establishment of science education at Oxford, partly for future medical students. Like Simon, Acland had a long-standing interest in pathology, and in preventive medicine.[1]

By the 1880s, however, important differences had emerged among the three men. Simon and Acland saw preventive medicine as the fundamental medical advance of their lifetime. In their opinion, therefore, the greatest hope for the immediate improvement of medical practices lay in studies of disease—that is, in pathology rather than in physiology. Neither Simon or Acland was a laboratory researcher. Neither one fully understood the application of the experimental method to physiology or pathology. Their views had been formed in an earlier era in which more descriptive, cataloguing methods had been the rule in the life sciences.

Dismayed at the erosion of the physician-humanist ideal that had previously prevailed at Oxford, Acland was also uncomfortable with vivisection — the most important method for physiology, in Burdon Sanderson's opinion. Being more supportive of the research ideal in science, Simon favored the use of animals in pathological research. But in 1890 even he backed Acland's public criticisms of the reforms in medical education at Oxford.[2]

These three men's conceptions of scientific medicine were multifaceted and cut across many disciplinary boundaries. Each man had different priorities. Their differences were emblematic of the disagreements and confusions about aims among proponents of scientific medicine.

In the 1890s the antivivisectionists, who rejected experimentalism, were less influential than in the two preceding decades. But among the advocates of scientific medicine schisms emerged, perhaps because they felt they no longer needed to present a united front against a strong enemy. At Oxford ongoing disputes among the three biology programs illustrated that there was no consensus on the importance of the emerging biological disciplines to medicine. Underlying these turf battles were competing models of physiology — a more British "biological" model and the more Continental "physical" model that Burdon Sanderson espoused.

Space, both literally and in the curriculum, was made for physiology and pathology at Oxford. But many in the medical profession were not convinced of the utility and relevance of experimental medicine to practice and did not agree that sciences such as physiology should be the foundation of the medical curriculum. The mixed reaction within the medical community to the germ theory and the establishment of the field of bacteriology — the premier contribution of laboratory research to medical practice — reflected this ambivalence.

It was bacteriology, however, which most dramatically illustrated that conditions for laboratory research in Britain — in terms of facilities, positions, and funding — were substantially different than they had been in earlier in the century. A child of pathology, bacteriology was so vigorous by the 1890s that its researchers had almost taken over the London Pathological Society, if the percentage of its journal devoted to their publications is a guide. It was the very dynamism of disciplinary boundaries which led to antagonisms across those moving borders, and researchers found themselves jostling for funding and influence, as detailed in verse in the *Oxford Magazine* in 1892:

The Disputes within the Biology Department

The person with a taste for Greek his taste may gratify,
For modern study's advocates have other fish to fry:
The undissected frog may hop besides Cherwell's brim—
For Science has no time just now for vivisecting him.

In every Term 'tis still the same—they crack each other's crowns:
Professor Brown wants Jones's men,—Professor Jones wants Brown's:
While great Professor Robinson with neither side agree—
It is not that he wants the men—but much he wants their fees.[3]

The author of these lines was lampooning the Oxford life science departments. The first stanza alluded to problems that the physiology department
had faced: the antivivisectionists and the larger body of traditionalists within
Oxford. The second stanza referred to Burdon Sanderson and to E. Ray
Lankester, the Linacre Professor of Comparative Anatomy, who were fighting over students using degree statutes as their weapons. The "great Professor
Robinson" was likely the long-serving Regius Professor of Medicine, Acland.

Rooted in a competition for students among the three biology departments (physiology, morphology, and botany), these disputes had evolved out
of the university examination structure. A degree candidate who planned to
take a Natural Sciences Honours School in a biological subject first took a
preliminary examination in morphology, physiology, and botany. But since
1886 their final degree examination, the Natural Sciences Honours School,
would be only in one of the three subjects. Thus, at Oxford the development of the medical school, whose students were encouraged to specialize in
physiology, meant the devastation of student numbers in the other biological
branches. Janet Howarth tabulated the results:

*Natural Science Honours Graduates in the Biology
Departments of Oxford, 1886–1900*

	Physiology	Morphology	Botany
1886–89	29	12	1
1890–94	50	6	5
1895–99	72	12	3

In contrast, at Cambridge degree candidates chose two fields; thus, although there, too, physiology was the usual choice for medical students, the other biological sciences benefited from the medical school.[4]

The appointment of Lankester as Linacre Professor in 1890 had exacerbated the situation. He had been selected, with Burdon Sanderson's support, despite worries about the suitability of his character. As Lankester put it succinctly to Huxley: "I am trying to get back to Oxford to take poor Moseley's place. There is some difficulty as they do not all rightly value the blessing of having someone among them to prosecute the Vice Chancellor and otherwise make things lively."[5]

Almost immediately, Lankester pressed for a degree system like Cambridge's to encourage more students to study comparative anatomy (morphology) and botany. Burdon Sanderson and the physical scientists opposed the proposal because they believed that having students study more than one honors subject would result in lower standards; not coincidentally, the examination structure was advantageous for their disciplines.[6] The proposals were unsuccessful.

The controversy erupted again when it was proposed, in yet another attempt to shorten the medical program, that medical students take only one combined preliminary exam in botany and morphology. In response Lankester mounted a virulent and public counteroffensive—he attempted to ban undergraduates from studying human anatomy[7] and to force physiology students to be tested in morphology in their final examination. Rather effectively, Burdon Sanderson could quote Lankester to support his own position, because Lankester in the 1870s had been critical of Acland and George Rolleston (then Linacre Professor) and publicly espoused the cause of a medical school at Oxford. Now Lankester found himself recanting; his program could not compete against physiology and the medical school for students.[8]

Burdon Sanderson won out against changes in the statutes, but he alienated Lankester, who wrote to Acland: "I am afraid of a Committee of the 'Biological professor[s].' I know you to be a naturalist and wide-minded philosopher." Ironically, he was now allied with Acland, his sworn enemy in the 1870s. Lankester continued: "But Sanderson is not: nor would our *new* Professor of Human Anatomy be so. Hereafter on such a committee the narrow technical medical spirit would prevail & I should be in a minority of one. I

Figure 8.1. John Burdon Sanderson, 1894,
Bodleian Library, MS, Don d. 14, fol. 25,
Sarah Angelina Acland.

should prefer a committee chosen *ad hoc.*"[9] Lankester was not really interested in medical training; what he wanted was to establish a research discipline, and for him having medical students was a means to an end. Burdon Sanderson, in contrast, remained committed to the reform of the practice of medicine, despite his fervent desire to establish a research program.

In the background medical practitioners themselves remained unconvinced of the relevance of laboratory medicine to practice, and they were likely not entirely sympathetic to Burdon Sanderson's predicament. The responses to Acland's 1890 article "Oxford and Modern Medicine," which attacked the recent changes at the university, revealed these critical sentiments. Acland received numerous letters of congratulation. In the article he had reiterated that Oxford should educate small numbers of elite practitioners by first having them undergo a general literary education together with some broad training in the sciences; they should then go on to metropolitan medical schools to complete their training. In general his correspondents expressed dissatisfaction with the increasing emphasis on the preliminary sciences in medical education. The fear was that science was squeezing out general culture and wasting time better spent in professional training (meaning practical clinical education).[10]

The public dispute between Burdon Sanderson and Lankester brought disrepute to Oxford's biology departments as each attempted to gain supporters from the British scientific community. William Church stated, "It is amusing, though I fear detrimental to Oxford, that each seems bent on cutting into the claims of the other on the students' time."[11] This fight underscored Burdon Sanderson's inability to operate effectively at Oxford. The physiology department and the medical school required ongoing changes in the statutes for which there was sure to be opposition from outside the science departments and outside the university. In the hostile environment at Oxford the life science departments needed to stand together. Lankester, with his famously prickly personality, was a problem in his own right, but Burdon Sanderson's refusal to support any change to ameliorate the declining numbers in botany and morphology was equally to blame. Yet the controversy between Lankester and Burdon Sanderson was also due to a fundamental disagreement about the nature of physiology.

Evolution and the Place of Physiology in the Biological Sciences

For Burdon Sanderson physiology should be allied with chemistry and physics, not with zoology and botany. He stated unequivocally: "I do not think it desirable that students in my subject should spend more time than they do at present in the preliminary study of biology. . . . But I am most desirous to have men well trained in Physics & Chemistry."[12] His refusal to support more time spent on comparative anatomy reflected his belief that it was relatively unimportant for physiology. Lankester, in contrast, believed that it was important to develop a "biological attitude" in physiology. Physiologists without a biological attitude, he believed (paraphrasing Professor Max Verworm of Jena), were "too much set on mere physical and chemical determinations without aiming clearly at the real problems of Life."[13]

Burdon Sanderson's vision of physiology brought him into conflict with

British contemporaries. Historians Gerald L. Geison and Richard D. French have described the British national style of physiological research in this period as "evolutionary."[14] They contrasted this British style with that of Continental researchers, who were less influenced by evolution. Their conclusion was based on a group of researchers centered around Foster and Huxley.

Without question Burdon Sanderson accepted Darwin's theory of evolution. He considered it uncontroversial; he frequently used its status to bolster experimental physiology in the public eye. In his Royal Institution lectures he subtly and directly associated physiology with Darwinian evolution; on one occasion he explained that Johannes Müller (1801–58), the German physiologist, "no less than Darwin, by his influence on his successors was the beginner of a new era."[15] But, despite this connection of Darwin with physiology, he did not consider evolution terribly relevant to the science. He demarcated the science of physiology from the "other great branch of Biology, the Science of Living Beings," which began "as we now know it" with "the appearance of the *Origin of Species.*"[16]

The differences between Burdon Sanderson and the Huxley group should not be exaggerated. Of course, the Huxley group was well aware of the achievements of German physiology: Lankester, for example, had studied with Carl Ludwig (1816–95) at Leipzig. Similarly, Burdon Sanderson was influenced by evolutionary ideas—as French noted, his Venus's-flytrap experiments were suggested by Charles Darwin himself. Nonetheless, for the Huxley group physiology belonged to the discipline of biology. Huxley, in particular, articulated the view that all living organisms—plants and animals—had common structures, functions, and evolutionary histories.[17] Geison's and French's conclusions were apt for the group of men they described. But Burdon Sanderson was on the periphery of that group.[18]

For Burdon Sanderson physiology should be built on physical and chemical knowledge. In his opinion physiologists should look abroad, particularly to German universities, for their models. Burdon Sanderson's vision offered an alternate model of physiology to the British research community.

Pathology versus Physiology

Another ongoing source of irritation between Burdon Sanderson and the Huxley camp was his serious commitment to research pathology. Despite their portrayal of him as aberrant, others in his generation pursued eclectic research careers. His French colleague Chauveau published on comparative anatomy, contagion, and cardiac function. Similarly, Panum, "the founder of Danish physiology," also wrote a book on a measles epidemic.[19]

The new laboratory-based experimental pathology and physiology were the intellectual heirs of the pathological anatomy of the turn of the century. To generalize: physiologists studied normal function, using vivisection; pathologists, inspired by Virchow's cellular pathology model, studied disease phenomena, such as inflammation, at the cellular level. In reality the separation of the two fields was untidy. Pathologists and physiologists both used animals, and they shared equipment, techniques, and laboratories; sometimes pathologists were physiologists, and vice versa. Burdon Sanderson's career embodied the porous and dynamic nature of the boundary between physiology and pathology which existed for much of the nineteenth century.

To a large extent the separation of the two fields has been written backwards into the literature; it is an artifact of the present. Jacalyn Duffin pointed out that Laennec had been eliminated from the history of physiology, although he described himself as contributing to pathological anatomy and physiology. In a similar fashion the contributions of men like Burdon Sanderson, Chauveau, and Panum are frequently truncated; parts of their research lives appear in general histories of physiology, others in histories of cardiology, bacteriology, or veterinary science.[20]

In the British context the separation of physiology and pathology reflects the views of the Huxley group, whose adherents aimed to establish physiology, for all intents and purposes, separately from medicine. Huxley went so far as to argue that pathology itself should be "a pure science of medicine . . . which has no more necessary subservience to practical ends than has zoology or botany."[21]

After the Germ Theory

The tensions in the medical community deepened in reaction to the now well-established field of bacteriology. No practitioner would argue that laboratories were entirely irrelevant to the practice of medicine; however, doubts about the research interests of bacteriologists were expressed.[22] These were evident in 1891, when Burdon Sanderson delivered the Croonian Lectures on the "Progress of Discovery Relating to the Origin and Nature of Infectious Diseases." The clear hero of his story was Robert Koch (1843–1910), who had solved the two great problems of the germ theory, specificity and causality, with his techniques for isolating and cultivating "microphytes"— the term Burdon Sanderson used for these infectious microorganisms. First, Koch had grown the *Bacillus anthracis* outside of the body and described its life history, in the process establishing beyond all doubt its causal relationship with anthrax. Later Koch had developed the technique of cultivating microphytes on a solid media, which allowed for their sorting and separation.[23]

Despite Koch's triumphs, Burdon Sanderson thought that the relation between microorganisms and the process of inflammation was still not fully elucidated.[24] Following the German rather than the French school of bacteriology, Burdon Sanderson argued from the experimental literature—which had increased enormously in size since the 1880s—that "the proximate cause of inflammation is always chemical."[25] And, when he discussed the ability of the organism to arrest and destroy microphytes, he stated that this function resided "both in the blood and in the tissues" and "both in organised protoplasm and intercellular liquid, and consequently the *"inhibitory or destructive action is chemical."*[26]

Here Burdon Sanderson challenged Elie Metchnikoff's (1845–1916) theory that phagocytes destroyed microphytes within the body by incorporating them into their own cells. He contended that this theory had not been proven. There was too much contrary evidence; after all, as he explained, *"[All disease-producing microphytes] are sentenced to death. The question is, Is death always preceded by imprisonment?"*[27] The answer was an unequivocal no.[28]

In Burdon Sanderson's opinion the underdetermination of Metchnikoff's theory was symptomatic of the new field of bacteriology, and he contended that bacteriologists had largely abandoned medical practitioners. Although pathology had spawned bacteriology, the relationship between the two was unclear. Would bacteriology take over pathology? Or would it become a separate discipline, a kind of morphology of microorganisms?[29]

The separation between clinicians and bacteriologists was illustrated by Arthur Bower Griffiths, a nonmedically trained bacteriologist, in the preface to his 1891 book, *Researches on Micro-Organisms*. Griffiths wrote, "Although this work, in a great measure, has a bearing on the *treatment* of certain infectious diseases, I may say that I have had no desire to trespass in the domain of the physician."[30]

Burdon Sanderson emphasized that pathologists should not be interested in these organisms for their own sake but because of their abilities as "mischief-makers." Bacteriologists all too often were merely intent on making "splendid discoveries" that "so engrossed the attention of all that it was for a time forgotten" that the real focus should be not the microbes themselves but "their disease-producing powers."[31] Some of these objections were doubtless jealous jabs at the empire building of Pasteur and other bacteriologists. Burdon Sanderson had a long-standing dislike of Pasteur, whom he felt had never acknowledged his own anthrax research.[32] But he sincerely believed that bacteriologists were on the wrong path. While his specific criticisms may have been somewhat idiosyncratic, Burdon Sanderson's dismay with developments in the new field of bacteriology reflected the disappointment of those non-experimenters who had expected a revolution in therapy to come quickly in the wake of the germ theory.

Without question, specific microorganisms sometimes were seen to cause certain diseases; in that sense the germ theory was accepted. Koch's research, embodied in his famous postulates of 1880, had closed off debate on that issue. But this did not mean that all infectious diseases were caused by germs. More important, this closure had not directly resulted in any new therapies. Thus, much contested territory remained: Why did some people exposed to a bacteria remain healthy? Why did some fall ill, only to recover? Germ theory skeptics turned increasingly away from bacteriology and toward the body's response to infection—in a word, immunology—in the hope of finding the elusive cure (or preventive measures) for infectious diseases.

Burdon Sanderson was even more blunt in his private criticisms of bacteriology. A young aspiring researcher, R. H. Bremridge, wrote to thank him for his advice and promised: "I will always bear in mind what you said about Bacteriology. Everything now seems to find a supposed explanation in the microbe, and I can quite realise the fascination it may exercise over a beginner."[33]

The Corner Turned

In 1895 Burdon Sanderson became Regius Professor of Medicine. Being a personal triumph, his ascension to this very prestigious clinical post also proved that the research ideal in medicine had arrived in Britain. From this point on, he devoted more time to his family circle, making frequent visits to the Haldanes in Scotland, and his trips to the Continent were more social than professional. He was, however, not exactly retired.

His predecessor Acland had maintained a large practice and an interest in the prevention of medicine. No longer a clinician, as Regius Professor, Burdon Sanderson returned, in a more administrative capacity than before, to preventive medicine and pathology. At Oxford he initiated organized teaching in pathology, with an emphasis on practical or laboratory instruction. Through his efforts (which were often as frustrating as his experiences as Waynflete Professor) a readership in pathology was established, and a building designed specifically for pathology was constructed.

Burdon Sanderson was also closely involved in the establishment of the Lister Institute of Preventive Medicine. During an 1889 meeting in London, ostensibly organized to acknowledge Pasteur's work with over two hundred British rabies patients, it was decided that an institution similar to the Pasteur Institute or the Hygienic Institute in Berlin was needed in London. The British Institute of Preventive Medicine was incorporated in 1891 and was to include research into the study and prevention of infective animal and human diseases; instruction in preventive medicine; the treatment of persons suffering from infective diseases; and laboratories with scientific staff. In addition there were plans to arrange public lectures and demonstrations, publish transactions, and found a library. Renamed the Jenner Institute in 1898 to fulfill the conditions of a bequest, legal problems led to its rechristening in 1905 as the Lister Institute of Preventive Medicine.[34]

Despite these grandiose, albeit belated, plans for the Lister Institute, Britain no longer led the way in preventive medicine. Years after Burdon Sanderson's trip to Dantzig, he visited some German hospitals "and was astonished with the extraordinary contrast to his previous experience. The hospitals he saw then—one in particular at Halle—were extremely clean, airy, and apparently supplied with satisfactory nurses." He noticed particularly that "the antiseptic treatment was scrupulously carried out, whereas in England at the same time there were still many hospital physicians and surgeons who did not realize its importance, although this had been scientifically demonstrated by our own countryman, Lister."[35] Simon and Burdon Sanderson at the turn of the century mounted a public campaign (in the *Times*) to create a British system of sanitoria for the tubercular poor modeled on the German system.[36]

Britons had once smugly believed that their industrial advantage mirrored their own social and moral superiority; now they were becoming aware that their industrial might was being eclipsed.[37] Burdon Sanderson's reactions to sanitary conditions in the new Germany reflected the less optimistic tone of the late Victorian period. As he noted, the British public health campaign was no longer unique. Ironically, despite this pessimism, the actual health of Britons had greatly improved during the late nineteenth century. The worries of the period reflected the more intractable nature of the problems that remained and the difficulty and expense of producing what would likely be relatively small improvements in British living conditions.

The pessimism of the era should not be exaggerated. Burdon Sanderson remained convinced that advances in medical treatment were just ahead. He admitted that in the recent past pathologists had been too quick to hope for easy solutions, but he was certain that the corner had finally been turned. The latest discoveries about the ability of organisms to protect themselves against infection were "the beginning of a new progress," and he asserted confidently that it required no power of prophecy to see that "we are just now on the threshold of discoveries such as will surpass in their practical value . . . [and] scientific importance, all that have preceded." He had concluded his 1891 lecture by saying, "Happy are they who are young enough to hope to participate in the realization of so encouraging a prospect!"[38]

In 1899 Burdon Sanderson was made a baronet, and Michael Foster was knighted. Their dual honor was hailed by the physiological community. L. E. Shore wrote from Cambridge, "On Saturday we were all rejoicing in the laboratory at the honours given for Physiology first of all to you as chief of English Physiologists and then to Foster our own Professor."[39] Foster himself wrote to Sanderson, "Like you I don't care for these things—but we may both congratulate each other that our common science has received such marked recognition & it is above all things nice that *we should be* run in together." He continued, "Either common rumour as to the 'old lady's' prejudices are wrong or she must have screwed up her face when she accepted Salisbury's two suggestions."[40] The honors bestowed on Foster and Burdon Sanderson by Queen Victoria, who was reputed to be an antivivisectionist herself, further underlined the public triumph of experimental medicine over the antivivisectionists.

The Medical Community and the Ideals of Science

By the turn of the century experimental pathology was less contentious. In his 1902 presidential address to the London Pathological Society, Burdon Sanderson's tone was happier than that of his Croonian lecture nine years earlier. Although he emphasized, as always, that there was much hard work ahead, the future of experimental medicine which Burdon Sanderson had often predicted had finally arrived. Speaking of ongoing changes in beliefs about inflammation and infection, he stated, "Happily the change does not depend, as was wont to be the case at former periods in the history of medicine, on a swing of the pendulum of medical opinion, but on the real progress of experimental investigation."[41]

In the larger medical community many continued to oppose the ideals of a scientific medicine. But there had been a significant shift: it was increasingly obvious that this opposition was a reaction to the successful establishment of the new laboratory medical sciences. Divisions within the medical community were most evident in continuing disputes over the weighting of elements of the medical curricula. In 1902 C. S. Sherrington (1852–1952) wrote to Burdon Sanderson to ask his assistance in the fight against the medical curriculum of the new Liverpool University.[42] The dean, Andrew Melville Paterson, also holder of the chair of anatomy, had pushed through a curriculum that reduced by half the time devoted to physiology and by a third the time devoted to pathology. His action was partly a selfish defense of his own subject, Sherrington explained; "It is held that Physiology must 'cede to the more practically important subject, Human Anatomy' which competes with it for the student's time."

But, above all, the dean's move was a frontal attack on laboratory sciences, as "he fears that the whole tendency of medical education is going from the 'sound subjects to pseudo science.'" Sherrington emphasized that the dean traced "this evil in root to physiology, which in turn infects pathology. In fact his attempt—supported by a number of the Medical Faculty to paralyze the Physiology school, is a counterblast to the new Laboratory & post in Chemical Physiology & Pathology." Physiology and pathology were under attack because of their success and high public profile. The other professors were jealous of the new laboratory sciences—especially when outside donors wanted to fund new laboratories for them.

Sherrington's campaign was successful, and the Liverpool medical curriculum remained unchanged.[43] Nonetheless, the dean's skepticism about the laboratory ideal for science was shared—after all, the initial proposal had been passed by the entire faculty. Sherrington tried to suggest that it was only old fogeys who had supported the dean, but even he had to admit that there was wider support. Moreover, there was a second model of medical education emerging in Britain which emphasized professional training over scientific. Sherrington explained that "the example of Edinburgh & especially of the new University of Birmingham—where Physiology is even further reduced—are held up by our Dean as the ideals." The place of some laboratory sciences in the medical curriculum and medical school was not entirely secure.

Scientific Medicines

The trajectory of Burdon Sanderson's career demonstrates that the meaning of *scientific medicine* was and, I would argue, still is both unstable and multifaceted: sometimes it included the kind of work that medical officers of health did—cleaning canals and testing milk—other times experimental investigations of cows or puppy dogs, involving measuring instruments, microscopes, and microbes, and encompassed electrophysiological experiments on Venus's-flytraps.

Scientific medicine has always had its supporters and detractors. The real question should be what kind of activity was supported, when, and by whom. In Victorian Britain not everyone supported laboratory research, either physiological or pathological. In general physiology was supported because of its association with pathology and therapeutics, that is, the study of illnesses and their potential treatments. By the turn of the twentieth century a research infrastructure for laboratory medicine, and in particular experi-

Figure 8.2. Vanity Fair cartoon of
Burdon Sanderson, 1894,
Queen's Medical Photography.

mental physiology and pathology, had been established in Britain — clear evidence of support. But there were always those who felt that other sciences, such as hygiene or the clinical sciences, were more significant, pertinent, and informative.

Most important, the desire for coherence, for one scientific medicine, or an easily defined progression through the nineteenth century from pathological anatomy to laboratory medicine at the end is the result of later disciplinary divisions — and our desire for a neat story. The Victorians, unsurprisingly, did not see the world as we do. Where we see differences, they saw a coherent whole. Where we see coherency by relating the past to our present, they saw differences. Since they did not know how the story was going to turn out, they did not divide bodies of knowledge as we do. At no point (not even today) can we point to one Victorian activity and say that it defines scientific medicine. The meaning and boundaries of medical science were and continue to be constantly debated and negotiated.

Epilogue

Throughout his life Burdon Sanderson worked exceedingly hard to establish research pathology and physiology. He had aimed both to make discoveries and to create a place for the two fields in British universities, and his efforts were a success. Physiology and pathology both existed as research disciplines in Britain; their domains were more precisely defined and distinct from each other. Burdon Sanderson had lived long enough to see a significant shift in opinion, and he no longer needed to justify the role of the laboratory and experiment in medical education and therapeutic change.

But Burdon Sanderson never dwelt on his successes. As many contemporaries noted, he was self-critical and unable to take pleasure in his achievements or even relax—he never saw the work that had been done, only what remained to do. His self-criticism and incessant activity were the most lasting remnants of his early Evangelical beliefs. Burdon Sanderson never effectively cut back his work schedule, even when it threatened his health. Despite the many honors he received and his personal happiness, he always felt that he was a failure. He had not written the great synthetic work on electrophysiology which he had once planned.[44] Worse, he had never made a great discovery in physiology, and his Venus's-flytrap experiments were widely questioned. As George Eliot wrote of Lydgate, "He always regarded himself as a failure: he had not done what he once meant to do."

Outsiders saw a different picture. Sir John Burdon-Sanderson, Regius Professor of Medicine, had achieved status for himself and public recognition for experimental medicine. Not only had he managed to make a career as a research scientist—still an uncommon achievement in his era; he was also a pivotal figure in establishing laboratory research, both physiological and pathological, as a basis of medical education in Britain. His own pathological research was first-rate and gained him an enduring reputation, and he helped bring British medical researchers into the international community.

In 1904 Burdon Sanderson resigned the Regius Professorship of Medicine due to his failing health. He continued to weaken; with Ghetal by his side, John died peacefully on 23 November 1905.[45]

Abbreviations

JBS	John Burdon Sanderson
GBS	Ghetal Burdon Sanderson
Memoir	Ghetal Burdon Sanderson, *Sir John Burdon Sanderson: A Memoir,* completed after her death by J. S. and E. S. Haldane (Oxford: Clarendon Press, 1911).
Bodl.	The Bodleian Library, Oxford University
EUL	Edinburgh University Library
GLRO	Greater London Record Office
ICL	College Archives, Imperial College London
NLM	National Library of Medicine, Bethesda, Maryland
NLS	National Library of Scotland
PRO	Public Record Office, Kew, England
UBC	Charles Woodward Memorial Room, Woodward Library, University of British Columbia, Canada
UCL	The Library, University College London
WL	Contemporary Medical Archives Centre, Wellcome History of Medicine Library, London

Appendix

Researchers Associated with Burdon Sanderson
in Britain, with Selected Archival Sources

Sources

(*1*)   Edward Sharpey-Schafer, *History of the Physiological Society during Its First Fifty Years, 1876–1926* (Cambridge: Cambridge University Press, 1927).

(*2*)   W. J. O'Connor, *Founders of British Physiology: A Biographical Dictionary, 1820–1885* (Manchester: Manchester University Press, 1988).

(*3*)   Gerald L. Geison, *Michael Foster and the Cambridge School of Physiology: The Scientific Enterprise in Victorian Society* (Princeton: Princeton University Press, 1978).

(*4*)   Richard D. French, "Darwin and the Physiologists, or the Medusae and Modern Cardiology," *Journal of the History of Biology* 3 (1970): 253–74.

(*5*)   Steven Waite Sturdy, "A Co-Ordinated Whole: The Life and Work of John Scott Haldane" (Ph.D. diss., University of Edinburgh, 1987).

1. *William Maddock Bayliss* (1860–1924): student of JBS, researched at University College; ended his career as professor of general physiology, University of London. (*1*), 79–80; (*2*), 140. Bayliss to JBS, 3 June 1899, fols. 32–33, MS ADD 179/8, UCL.

2. *Charles Beevor* (1854–1908): bought instruments for JBS in France and German in early 1880s; in 1883 clinician at the National Hospital for the Paralysed and Epileptic researcher with Horsley (see entry) at the Brown Institution; also worked with Gotch and JBS in the laboratory; became president of the Neurological Society in 1907. (*1*), 83, 85. Beevor to JBS,

20 May 1883, fols. 80–81; 2 July 1883, fols. 82–83, MS20030, NLS. GBS, notes, bk. 3, 1886, fols. 68,71, Burdon Sanderson Collection, UBC.

3. *John Rose Bradford* (1863–1935): worked at the University College laboratory; became President of the Royal College of Physicians. (*1*), 79; (*2*), 140.

4. *Thomas Lauder Brunton* (1844–1916): in 1860s was personal assistant to JBS; in the 1870s worked with JBS at the Brown Institution; became noted pharmacological researcher; then devoted all his time to clinical practice. (*1*), 31. Ferrier to GBS, 13 December 1907, MS ADD 179/93, UCL.

5. *George Alfred Buckmaster:* pupil of JBS at Oxford; from 1927 professor of physiology at Bristol University. (*1*), 77; (*2*), 145. Buckmaster to JBS, 20 January 1895, fol. 37, MS ADD 179/7, UCL.

6. *John Theodore Cash:* worked with JBS in 1886; later professor of Materia Medica at Aberdeen. (*1*), 59. GBS, notes, bk. 3, 1886, fols. 68, 71–72, Burdon Sanderson Collection, UBC.

7. *F. A. Dixey:* studied medicine at University College Hospital; later histology lecturer at Oxford. (*1*), 82–83. JBS to Schäfer, 25 January 1885, 179/13, and 27 January 1885, PP/ESS/B8/8, WL. Dixey to JBS, 17 August 1886, fols. 72–73; 20 August 1886, fols. 74–75, MS ADD 179/5, UCL.

8. *David Ferrier* (1843–1928): worked with JBS at the Brown Institution in the 1870s. (*1*), 8; (*2*), 190. Ferrier to GBS, 13 December 1907, MS ADD 179/93, UCL.

9. *Walter Holbrook Gaskell* (1847–1914): worked at the University College laboratory in the early 1870s with JBS; then physiologist at Cambridge. (*1*), 33–34; (*2*), 174; (*3*); (*4*). Schäfer to JBS, 17 May 1872, fols. 33–34, MS20501, UCL. GBS, notes, bk. 2, fol. 140, Burdon Sanderson Collection, UBC.

10. *Francis Gotch* (1853–1913): Sharpey Scholar in 1881 at University College; JBS's assistant at Oxford; succeeded JBS as the Waynflete Professor of Physiology at Oxford. (*1*), 66, 151; (*2*), 140, 159. Gotch to JBS, 3 April 1883, fols. 137–40; 6 April 1883, fols. 141–44; 13 April 1883, fols. 14–18, MS ADD 179/4, UCL; and 8 May 1892, fols. 79–82, MS ADD 179/6, UCL.

11. *J. S. Haldane* (1860–1936): JBS's nephew; university demonstrator in physiology at Oxford. (*1*), 84; (*2*), 145; (*5*).

12. *Leonard Hill* (b. 1866): Hill did research in the Oxford physiology

laboratory; went on to a career in physiology in London. (*1*), 97; (*2*), 145. Hill to JBS, 17 February 1894, fols. 159–60; 26 August 1894, fols. 163–64, MS20501, NLS.

13. *Victor Horsley* (1857–1916): worked at University College laboratory in the early 1880s; professor-superintendent of the Brown Institution from 1884–90. Stephen Paget, *Sir Victor Horsley: A Study of His Life and Work* (London: Constable and Co., 1919). (*1*), 77, 155; (*2*), 140. Horsley to JBS, 27 April 1887, fols. 82–85; 27 April 1887, fols. 86–87, MS ADD 179/5, UCL; and 10 November 1896, fols. 132–33, MS ADD 179/7, UCL.

14. *A. F. S. Kent* (1863–1932): worked with JBS at Oxford; became physiological demonstrator at Owen's College, Manchester, and physiology professor at Bristol University. (*1*), 85; (*2*), 145.

15. *E. E. Klein* (1844–1925): brought to London by JBS as assistant; prominent bacteriologist. (*1*), 31; (*2*), 155–57.

16. *Frederick W. Mott* (1854–1926): did research in the University College laboratory on bacteriology; became an eminent pathological neurologist. (*1*), 81. Mott to JBS, 7 June 1899, fols. 50–51, MS ADD 179/8, UCL.

17. *William North* (b. 1854): Sharpey Scholar at University College from 1879 to 1881. (*1*), 53; (*2*), 223, 159; (*3*), 373.

18. *William Osler* (1849–1919): worked in the laboratory at University College London in the early 1870s. (*2*), 140.

19. *F. J. M. Page* (1848–1907): JBS's longtime assistant in his electrophysiological research; became a lecturer in physics and chemistry to medical students in London. (*1*), 47; (*2*), 159–60

20. *Marcus Seymour Pembrey* (1868–1934): worked with JBS at Oxford; in 1927 he was the professor of physiology at Guy's Hospital. (*1*), 104–5; (*2*), 145, Pembrey to JBS, 3 June 1899, fol. 1, MS ADD 179/8, UCL.

21. *Sydney Ringer* (1835–1910): used the laboratory at University College from 1875 to 1895. (*1*), 55; (*2*), 154.

22. *George John Romanes* (1848–94): frequent visitor at the Burdon Sanderson home until the end of his life; he had worked with JBS in London from 1874 to 1876. (*2*), 249–51; (*3*); (*4*). GBS, notes, bk. 2, fols. 140, 186, WL-UBC. Francis Galton to JBS, 29 September 1894, fol. 167–68; 21 October 1894, fols. 169–70, MS ADD 179/6, BSC, UCL. Herbert Spencer to JBS, 10 November 1894, fols. 175–76, MS ADD 179/6, BSC, UCL.

23. *James Lorrain Smith* (1862–1931): worked in the Oxford physiology laboratory; became professor of pathology at Belfast, Manchester, and Edinburgh in turn. (*1*), 98; (2), 145; (*3*), 374.

24. *Edward Bradford Titchener* (1867–1927): spent a year in the physiology laboratory at Oxford; in 1895 was appointed Sage Professor of Psychology at Cornell. Titchener to JBS, 21 June 1895, fol. 84, MS ADD 179/7, UCL.

25. *H. M. Vernon* (b. 1870): graduated from Oxford in physiology, became a physiological demonstrator at Oxford. (*1*), 117. Lankester to JBS, n.d., fols.198–99; n.d., fols. 202–3, MS20031, NLS. Vernon to JBS, 10 May 1894, fols. 199–200, MS20501, NLS.

26. *Augustus D. Waller* (1856–1922): received a British Medical Association grant to study with JBS in the early 1880s at University College; researcher into the electrical activity in nerve and muscle; made a career as a lecturer in physiology in London. (*1*), 53; (2), 215–17. Waller to JBS, 22 June 1890, fols. 12–13, MS ADD 179/6, UCL.

Notes

Introduction

1. This summary is derived from a vast literature in the history of physiology, pathology, and other medical sciences. On the transformations in the modes of medical sciences during the nineteenth century, see the work of John Pickstone, for example, "The Biographical and the Analytical: Towards a Historical Model of Science and Practice in Modern Medicine," in *Medicine and Change: Innovation, Continuity, and Recurrence: Historical and Sociological Perspectives,* ed. Ilana Lowy et al. (Paris: INSERM and John Lowy, 1993). He drew on Nicholas Jewson, "The Disappearance of the Sick Man from Medical Cosmology," *Sociology* 10 (1976): 225–44; and Erwin H. Ackerknecht, *Medicine at the Paris Hospital, 1794–1848* (Baltimore: Johns Hopkins Press, 1967). On pathology, Russell Maulitz's work strongly influenced my analysis; see *Morbid Appearances: The Anatomy of Pathology in the Early Nineteenth Century* (Cambridge: Cambridge University Press, 1987); and his earlier articles, including "Rudolf Virchow, Julius Cohnheim and the Program of Pathology," *Bulletin of the History of Medicine* 52 (1978): 162–68; and "Pathology," 122–42; Ronald Numbers, ed., *The Education of American Physicians: Historical Essays* (Berkeley: University of California Press, 1980). On physiology, see W. Bruce Fye, *The Development of American Physiology: Scientific Medicine in the Nineteenth Century* (Baltimore: Johns Hopkins University Press, 1987); Gerald L. Geison, *Michael Foster and the Cambridge School of Physiology* (Princeton: Princeton University Press, 1978); and William Coleman and Frederic L. Holmes, eds., *The Investigative Enterprise: Experimental Physiology in Nineteenth-Century Medicine* (Berkeley: University of California Press, 1989). See also Jacalyn Duffin, *To See with a Better Eye: A Life of R. T. H. Laennec* (Princeton: Princeton University Press, 1998), 286–94; and William F. Bynum, *Science and the Practice of Medicine in the Nineteenth Century* (Cambridge: Cambridge University Press, 1994).

2. Joan Perkin, *Women and Marriage in Nineteenth-Century England* (London: Routledge, 1989), discussed Victorian marriage; see esp. 264–69.

3. Jose Harris, *Private Lives, Public Spirit: Britain, 1870–1914* (London: Penguin, 1993), 150. There is a vast and contradictory literature on late-Victorian secularization. On Evangelicalism, see David Hempton, *The Religion of the People: Methodism and Popular Religion, 1750–1900* (London: Routledge, 1996); and Boyd Hilton, *The Age of Atonement: The Influence of Evangelicalism on Social and Economic Thought, 1795–*

1865 (Oxford: Clarendon Press, 1988). On secularization and religion more generally, see Alan D. Gilbert, *Religion and Society in Industrial England: Church, Chapel and Social Change, 1740–1914* (London: Longman, 1976); and Owen Chadwick, *Secularization of the European Mind in the Nineteenth Century* (Cambridge: Cambridge University Press, 1975).

4. Burton J. Bledstein, *The Culture of Professionalism* (New York: Norton, 1976); and Harold Perkin, *The Rise of Professional Society: England since 1880* (London: Routledge, 1989), provide an overview of professionalism. From a large literature on the history of the medical profession, see particularly M. Jeanne Peterson, *The Medical Profession in Mid-Victorian London* (Berkeley: University of California Press, 1978); Samuel E. D. Shortt, "Physicians, Science and Status: Issues in the Professionalization of Anglo-American Medicine in the Nineteenth Century," *Medical History* 27 (1983): 51–68; Matthew Ramsey, "The Politics of Professional Monopoly in Nineteenth-Century Medicine: The French Model and Its Rivals," *Professions and the French State, 1700–1900,* ed. Gerald L. Geison (Philadelphia: University of Pennsylvania Press, 1984), 225–305; Jose Parry and Noel Parry, *The Rise of the Medical Profession: A Study of Collective Social Mobility* (London: Croon Helm, 1976).

5. Perkin, *Rise of Professional Society,* xi–xii. On public health in Britain, see Anthony S. Wohl, *Endangered Lives: Public Health in Victorian Britain* (London: J. M. Dent and Sons, 1983); John M. Eyler, *Victorian Social Medicine: The Ideas and Methods of William Farr* (Baltimore: Johns Hopkins University Press, 1979); and Christopher Hamlin, *A Science of Impurity: Water Analysis in Nineteenth-Century Britain* (Berkeley: University of California Press, 1990).

6. Walter E. Houghton, *The Victorian Frame of Mind, 1830–1870* (New Haven: Yale University Press, 1957), 27–53; Gertrude Himmelfarb, *Marriage and Morals among the Victorians and Other Essays* (New York: Vintage Books, 1987); and Francis M. L. Thompson, *The Rise of Respectable Society: A Social History of Victorian Britain, 1830–1900* (Cambridge: Harvard University Press, 1988), all discuss Victorian optimism.

7. Houghton, *Victorian Frame of Mind,* 37. •

8. Newman, *Apologia pro Vita Sua,* quoted by Houghton, *Victorian Frame of Mind,* 41.

9. Royston Lambert, *Sir John Simon, 1816–1904, and English Social Administration* (London: MacGibbon and Kee, 1963) remains the classic account of Simon's life.

10. Frank Miller Turner, *Between Science and Religion: The Reaction to Scientific Naturalism in Late Victorian England* (New Haven: Yale University Press, 1974).

11. Geison, *Michael Foster,* 358–59.

12. John Servos, "Research Schools and Their Histories," in *Research Schools: Historical Reappraisals,* ed. Frederic L. Holmes and Gerald L. Geison, *Osiris,* 2d ser., 8 (1993): 3–15; Jack B. Morrell, "The Chemist Breeders: The Research Schools of Liebig and Thomson," *Ambix* 19 (1972): 1–46; and Gerald L. Geison, "Scientific Change, Emerging Specialties, and Research Schools," *History of Science* 19 (1981): 20–40.

13. Karl E. Rothschuh, *History of Physiology,* trans. Guenter Risse (Huntington, N.Y.: Robert E. Krieger Publishing Co., 1973), remains the classic overview of the field. Thomas D. Brock, *Robert Koch: A Life in Medicine and Bacteriology* (Madison: University

of Wisconsin Press, 1988), provides an overview of Koch's achievements. Bernard's own *Introduction to the Study of Experimental Medicine* is a compelling summary of his philosophy of research. On Pasteur, see Gerald L. Geison, *The Private Science of Louis Pasteur* (Princeton: Princeton University Press, 1995); Patrice Debré, *Louis Pasteur*, trans. Elborg Forster (first French ed., 1994; Baltimore: John Hopkins University Press, 1998); and Bruno Latour, *The Pasteurization of France*, trans. A. Sheridan Smith and J. Law (Cambridge: Harvard University Press, 1988).

14. Richard D. French, *Antivivisection and Medical Science in Victorian Society* (Princeton: Princeton University Press, 1975); and Nicolaas A. Rupke, ed., *Vivisection in Historical Perspective* (London: Routledge, 1990).

15. John Harley Warner, "Science in Medicine," *Osiris*, 2d ser., 1 (1985): 37–58; and "The History of Science and the Sciences of Medicine," *Osiris*, 2d ser., 10 (1995): 164–93, provide a survey of the literature. See also Gerald L. Geison, "Divided We Stand: Physiologists and Clinicians in the American Context," in *The Therapeutic Revolution: Essays in the Social History of American Medicine*, ed. Charles Rosenberg and Morris J. Vogel (Philadelphia: University of Pennsylvania Press, 1979), 67–90; and Christopher Lawrence, "Incommunicable Knowledge: Science, Technology and the Clinical Art in Britain, 1850–1914," *Journal of Contemporary History* 20 (1985): 503–20. For the important German context, see Arleen Marcia Tuchman, *Science, Medicine, and the State in Germany: The Case of Baden, 1815–1871* (Oxford: Oxford University Press, 1993).

16. Particularly relevant are the essays in Adele E. Clarke and Joan H. Fujimura, eds., *The Right Tools for the Job: At Work in Twentieth-Century Life Sciences* (Princeton: Princeton University Press, 1992).

17. John Harley Warner, "The Fall and Rise of Professional Mystery: Epistemology, Authority and the Emergence of Laboratory Medicine in Nineteenth-Century America," in *The Laboratory Revolution in Medicine*, ed. Andrew Cunningham and Perry Williams, 110–41 (Cambridge University Press, Cambridge, 1992), 112. See also the discussion of Cambridge in Mark W. Weatherall, "Making Medicine Scientific: Empiricism, Rationality, and Quackery in Mid-Victorian Britain," *Social History of Medicine* 9 (1996): 175–94.

18. Bynum, *Science and the Practice of Medicine*, also argues that the features of the medicine of today were in place by World War I.

Chapter One: Choosing Medicine

1. George Henry Lewes to Alexander Main, 5 December 1872, in *The George Eliot Letters*, vol. 5: *1869–1871*, ed. Gordon S. Haight (New Haven: Yale University Press, 1955), 337–38. See also "Quarry for *Middlemarch*," in *Middlemarch: An Authoritative Text, Backgrounds, Reviews and Criticisms*, ed. Bert G. Hornback (New York: Norton, 1977), 607–44; and *George Eliot's Middlemarch Notebooks, A Transcription*, ed. John C. P. Pratt and Victor A. Neufeldt (Berkeley: University of California Press, 1979).

2. See Geison, *Michael Foster*, esp. chap. 1. Fye, *Development of American Physiology*, discusses the Americans.

3. The main published source for JBS's life is the *Memoir*. The standard obituaries are: Francis Gotch, *Proceedings of the Royal Society London* 79B (1907): iii–xviii; Francis Gotch, *Dictionary of National Biography*, supp. 2, no. 1 (1912): 267–69; Sir Arthur MacNalty, *Proceedings of the Royal Society of Medicine* (London) 47 (1954): 754–58. See also Gerald L. Geison, "John Scott Burdon-Sanderson," in *Dictionary of Scientific Biography*, ed. Charles Gillespie (New York: Charles Scribner's Sons, 1970-), vol. 2, 598–99; and W. J. O'Connor, *Founders of British Physiology: A Biographical Dictionary, 1820–1885* (Manchester: Manchester University Press, 1988), 141–46.

For a discussion of small landowners, see David Cannadine, *The Decline and Fall of the British Aristocracy* (New York: Anchor Books, 1990), 9; he made the point about the Eldons and Haldanes on 111, 209.

4. Gertrude Himmelfarb, "A Genealogy of Morals: From Clapham to Bloomsbury," *Marriage and Morals*, 24–25.

5. *Mary Elizabeth Haldane: A Record of a Hundred Years (1825–1925)*, ed. Elizabeth S. Haldane (London: Hodder and Stoughton, n.d.), see esp. 47, 55, 116. Thompson, *Rise of Respectable Society*, 125–27, discussed upper- and middle-class childhoods.

6. Frank Miller Turner, "The Victorian Crisis of Faith and the Faith That Was Lost," in *Victorian Faith in Crisis: Essays on Continuity and Change in Nineteenth-Century Religious Belief*, ed. Richard J. Helmstadter and Bernard Lightman, 9–38 (London: Macmillan, 1990), 20–21. See also Henry Colin Gray Matthew, *Gladstone, 1809–1874* (Oxford: Oxford University Press, 1988), 6–8; and the discussion of the "moral struggle" for Victorians, Houghton, *Victorian Frame of Mind*, 233–35. Ford K. Brown, *Fathers of the Victorians: The Age of Wilberforce* (Cambridge: Cambridge University Press, 1961), provides a summary of the Evangelical movement and the changes that it brought to British society (4–6).

7. Elizabeth Haldane, *Mary Elizabeth Haldane*, 57; and *Alton Locke, Tailor and Poet*, Charles Kingsley (1850), quoted by Houghton, *Victorian Frame of Mind*, 63–64.

8. Elizabeth Sanderson to JBS, 10 January 1838 [fragment], 3–4, MS ADD 179/97, UCL.

9. Hempton, *Religion of the People*, 55–56, outlines the complex religious boundaries of the period.

10. Elizabeth Haldane, *Mary Elizabeth Haldane*, 94.

11. [Unidentified parent of JBS], 26 June 1851, fol. 2, MS ADD 179/97, UCL.

12. Gilbert, *Religion and Society in Industrial England*, 75, 136; Brown, *Fathers of the Victorians*, 2–6; Houghton, *Victorian Frame of Mind*, 185–86; and Matthew, *Gladstone*, 8.

13. Richard Burdon Sanderson to JBS, [fragment] 1849 or 1850, MS ADD 179/97, UCL.

14. See John Harley Warner, " 'Exploring the Inner Labyrinths of Creation': Popular Microscopy in Nineteenth-Century America," *Journal of the History of Medicine and Allied Sciences* 37 (1982): 7–33, esp. 8–9, for a discussion of microscopy in this period.

15. JBS's course record from medical school (with instructors in parentheses): 1847–48: Anatomy and Practical Anatomy (Goodsir), Chemistry (Gregory), Botany

(Balfour), and he attended hospital; 1848–49: Surgery (Miller); Institutes of Medicine (Bennett); Anatomy and Practical Anatomy and Demonstrations (Goodsir); Botany (Balfour); Practical Pharmacy (Chapman); Practical Chemistry (Kemp); Clinical Surgery (Syme), and he attended hospital and did his dressership; 1849–50: Materia Medica (Christison); Surgery (Miller); Institutes of Medicine (Bennett); Anatomy and Practical Anatomy and Demonstrations (Goodsir); Practice of Medicine (William Pulteney Alison); Clinical Surgery (Syme), and he attended hospital; 1850–51: Medical Jurisprudence (Traill); Clinical Medicine (Alison and others); Dispensary Practice (Gillespie); Practical Pharmacy (Chapman); Pathology (Henderson); Practice of Medicine (Alison), and he did his clinical clerkship under Alison. Examinations, MS Da37, EUL.

16. For this discussion of the Edinburgh medical school, I am particularly indebted to Michael Barfoot, personal communication.

17. *Memoir*, 23. See also John D. Comrie, *History of Scottish Medicine*, 2d ed., 2 vols. (London: Bailliére, Tindall and Cox, 1932), 2:608.

18. Houghton, *Victorian Frame of Mind*, 237–38, discusses Evangelicalism and work.

19. *Memoir*, 29.

20. JBS to sister Jane, copy, n.d., probably 1848, fols. 30–32, MS ADD 179/97, UCL.

21. *Memoir*, 30–31.

22. R. G. C. Desmond, "John Hutton Balfour," *Dictionary of Scientific Biography*, 1:423.

23. Rudolf Virchow, *Cellular Pathology, as Based upon Physiological and Pathological Histology*, trans. F. Chance (London: John Churchill, 1860), quoted in L. Stephen Jacyna, "John Goodsir and the Making of Cellular Reality," *Journal of the History of Biology* 16 (1983): 75. See also David Heppel, "John Goodsir," *Dictionary of Scientific Biography*, 5:469–71; and Comrie, *Scottish Medicine*, 619–20.

24. Jacyna, "Goodsir and the Making of Cellular Reality," 80–82.

25. Ibid., 85–87. Goodsir, "Lectures on Comparative Anatomy" (1848), fols. 1–2, MSS Gen. 290, quoted by Jacyna, "Goodsir and the Making of Cellular Reality," 87–88.

26. For Goodsir's views, see Jacyna, "Goodsir and the Making of Cellular Reality," 88, 90, including the quotation cited from Goodsir, "Lectures on Comparative Anatomy" (1848), fols. 2–7, MSS Gen. 290, EUL, 88. For JBS's opinion, see *Memoir*, 23.

27. Jacyna, "Goodsir and the Making of Cellular Reality," 78–79.

28. Ibid., 98; and Heppel, "John Goodsir."

29. Jacyna, "Goodsir and the Making of Cellular Reality," 98–99; and Heppel," John Goodsir," 470.

30. Comrie, *Scottish Medicine*, 608; and John Harley Warner, "Therapeutic Explanation and the Edinburgh Bloodletting Controversy: Two Perspectives on the Medical Meaning of Science in the Mid-Nineteenth Century," *Medical History* 24 (1980): 248.

31. On Bennett's beliefs, see Warner, "Edinburgh Bloodletting Controversy," 249–

51. William Coleman, *Biology in the Nineteenth Century: Problems of Form, Function and Transformation* (Cambridge: Cambridge University Press, 1971), 23, discusses the cell theory.

32. JBS, 1901 dedication, *Memoir,* 24; and see also L. Stephen Jacyna, "Theory of Medicine, Science of Life: The Place of Physiology in the Edinburgh Medical Curriculum, 1790–1870," *Clio Medica* 30 (1995): 141–52.

33. *Memoir,* 24–25.

34. John Hughes Bennett Papers, Gen 2007, EUL, 1848, quoted by Warner, "Edinburgh Bloodletting," 241.

35. Warner, "Edinburgh Bloodlettting," 242, 247.

36. Ibid., 244–45, 251–53, 257–58; and John Harley Warner, *The Therapeutic Perspective: Medical Practice, Knowledge, and Identity in America, 1820–1885* (Cambridge: Harvard University Press, 1986), 217–19. See also Comrie, *Scottish Medicine,* 510–11 and, on Alison, 610–13. On the revival of bloodletting in the late eighteenth and early nineteenth centuries, see Guenter Risse, "The Renaissance of Bloodletting: A Chapter in Modern Therapeutics," *Journal of History of Medicine and Allied Sciences* 34 (1979): 3–22.

37. John Gray M'Kendrick, "Obituary: John Hughes Bennett," *British Medical Journal,* 1885, 475, quoted in Jacyna, "Theory of Medicine, Science of Life," 150. On Thomson's teaching of JBS, see *Memoir,* 39.

38. *Memoir,* 25.

39. This sentence is a paraphrase of Warner's comment about Bennett, "Edinburgh Bloodletting," 254.

40. *Memoir,* 27.

41. Ibid., 28.

42. John Scott Sanderson thesis, 1851, EUL.

43. L. Stephen Jacyna, "Robert Carswell and William Thomson at the Hôtel-Dieu of Lyons: Scottish Views of French Medicine," in *British Medicine in an Age of Reform,* ed. Roger French and Andrew Wear (London: Routledge, 1991), 110–35, discusses two members of the older generation. On the English and American pilgrims, see Russell C. Maulitz, "Channel Crossing: The Lure of French Pathology for English Medical Students, 1816–36," *Bulletin of the History of Medicine* 55, no. 4 (1981): 475–96; Maulitz, *Morbid Appearances;* and John Harley Warner, *Against the Spirit of System: The French Impulse in Nineteenth-Century American Medicine* (Princeton: Princeton University Press, 1998).

44. *Memoir,* 32.

45. F. W. Pavy to GBS, 10 June 1906, MS ADD 179/92, UCL.

46. *Memoir,* 36.

47. Gotch's obituary notice for the Royal Society, 3. Pavy and JBS's association with Bernard is noted in James Montrose Duncan Olmsted, *Claude Bernard, Physiologist* (New York: Harper and Brothers, 1938), 49; and *Memoir,* 33–34. On Bernard's legacy, see Mirko D. Grmek, *Le Legs de Claude Bernard* (France: Fayard, 1997).

48. John E. Lesch, *Science and Medicine in France: The Emergence of Experimental Physi-*

ology, 1790–1855 (Cambridge: Harvard University Press, 1984), 200, 209, 218. Frederic L. Holmes, *Claude Bernard and Animal Chemistry: The Emergence of a Scientist* (Cambridge: Harvard University Press, 1974), 246, 331. The quotation is from Olmsted, *Claude Bernard*, 48. The student was the American S. Weir Mitchell.

49. Olmsted, *Claude Bernard*, 41, 49. The first published lectures by Bernard were on blood, in 1853–54. Later Bernard regularly published his lectures, beginning with his course on "Experimental Physiology" at the Collège de France in the winter of 1854–55 (the course JBS had taken in 1851–52). Ibid., 55. See also JBS, "Notes on Bernard's Lectures," MS20504, NLS.

50. After moving to the Sorbonne in 1854, Bernard did not demonstrate for a decade, because of insufficient facilities. Olmsted, *Claude Bernard*, 52–53.

51. Ibid., 49.

52. Bernard to JBS, 25 June 1853, *Memoir*, 34–35; and fol. 2, MS ADD 179/1, UCL.

53. JBS, notes on Bernard's lectures, MS20504, NLS. On Bernard's surgical skills, see John Lesch, *Science and Medicine in France*. See also Claude Bernard, *Cahier de notes, 1850–1860*, ed. Mirko K. Grmek (Paris: Gallimard, 1965). In diary entries of 18 September 1871 (MS ADD 179/27, UCL) and 9 May 1872, JBS noted reading Bernard's lectures and meeting with Bernard on 29 January 1872 (MS ADD 179/28, UCL). The text with figures and suggested demonstrations appeared in 1873, in JBS, Edward E. Klein, Michael Foster, Thomas Lauder Brunton, *Handbook for the Physiological Laboratory*, ed. JBS (London: J. and A. Churchill, 1873).

54. Houghton, *Victorian Frame of Mind*, 218–38.

55. Matthew, *Gladstone*, 25.

56. Houghton, *Victorian Frame of Mind*, 220–22, 228.

57. JBS, 4 November 1863, "6–10 [P.M.] Nil Attendance on Parents," MS ADD 179/19, UCL; and 19 August 1864, "4.30–5 [P.M.] With Gh[etal] Kens[ington] Gardens," MS ADD 179/20, UCL.

58. Brown, *Fathers of the Victorians*, 6; and see the discussion of Evangelicalism in Noel Annan, *Leslie Stephen: The Godless Victorian* (London: Weidenfeld and Nicolson, 1984), 146–64.

59. "A Communication to the Harveian Society Entitled 'On Some Points in the Etiology of the Diseases of Infants under One Year,' Probably about 1857 . . . ," MS179/114, UCL.

60. Turner, "Victorian Crisis of Faith," 21; and see discussion on 20–34. Houghton, *Victorian Frame of Mind*, 48–50, discusses the "discomforts of belief."

61. Hilton, *Age of Atonement*, 268; and see 273–74. Houghton, *Victorian Frame of Mind*, 133, for the quotation from John Morley, 10–11. Himmelfarb, *Marriage and Morals*, 26.

62. Houghton, *Victorian Frame of Mind*, 169; and an opposing view, Frederick Gregory, "The Impact of Darwinian Evolution on Protestant Theology," in *God and Nature: Historical Essays on the Encounter between Christianity and Science*, ed. David C. Lindberg and Ronald L. Numbers (Berkeley: University of California Press, 1986), 369–90.

63. JBS to GBS, [14 June 1853], "Tuesday 3½ A.M." MS ADD 179/86, UCL.

64. On Victorian autobiography, see Linda H. Peterson, *Victorian Autobiography: The Tradition of Self-Interpretation* (New Haven: Yale University Press, 1986).

65. JBS, address to the Middlesex Hospital Medical School, 1 October 1868, MS ADD 179/114, UCL.

66. JBS, *On the Study of Physiology: Its Relation to Other Studies, and Its Use as a Preparation for That of Medicine* (Oxford: Parker and Co., 1883), 11–12, also reprinted in the *Memoir*.

67. Turner, *Between Science and Religion*, 8–37, and see particularly 11–12, 18.

68. For a discussion of positivism, agnosticism and Huxley's beliefs see Bernard Lightman, *The Origins of Agnosticism: Victorian Unbelief and the Limits of Knowledge* (Baltimore: Johns Hopkins University Press, 1987), 22–24. Ian Hacking, *Representing and Intervening: Introductory Topics in the Philosophy of Natural Science* (Cambridge: Cambridge University Press, 1983), 41–57, is a comprehensive review of positivism. The point about the similarity between scientific naturalism and Evangelicism is made by Turner, "Victorian Crisis of Faith," 17–20. See also Adrian Desmond, *Huxley: From Devil's Disciple to Evolution's High Priest*, originally published in 2 vols. (1994; rpt., Reading, Mass.: Addison-Wesley, 1997).

69. Turner, *Between Science and Religion*, 9 and see chap. 3 n. 29.

70. JBS later successfully asked Bernard and Wurtz for reference letters, *Memoir*, 35; and letter of 16 May 1853, fol. 1, MS ADD 179/1, UCL.

71. Richard Burdon Sanderson to JBS, [1852], MS ADD179/97, UCL.

Chapter Two: Medical Officer of Health

1. I assume GBS's mother was Helen Skirving Mowbray, the Rev. Herschell's first wife; his second wife was Esther Fuller Maitland. MS20033, NLS.

2. JBS to GBS, [February 1854], Edinburgh, MS ADD 179/86, UCL.

3. Thomas Ryan, secretary, St. Mary's Hospital, to GBS, 9 May 1907, MS ADD 179/93, UCL.

4. JBS to GBS, 1855, fol. 10, MS ADD 179/97, UCL.

5. Obituary notice of the Rev. Herschell's [1864], fol. 45, MS20033, NLS.

6. Christopher Lawrence, "Sanitary Reformers and the Medical Profession in Victorian England," in *Public Health: Proceedings of the Fifth International Symposium on the Comparative History of Medicine—East and West* (Tokyo: Saikon Publishing Co., 1981), 145–68, 153; and Margaret Pelling, *Cholera, Fever and English Medicine, 1825–1865* (Oxford: Oxford University Press, 1978), 231. On Chadwick and his era, see Christopher Hamlin, *Public Health and Social Justice in the Age of Chadwick: Britain, 1800–1854* (Cambridge: Cambridge University Press, 1998).

7. John Hargreaves Harley Williams, *A Century of Public Health in Britain, 1832–1929*, (London: A & C Black, Ltd., 1932), 271.

8. Lambert, *John Simon*, 104.

9. Quoted in ibid., 165.

10. Draft letter, fols. 4–5, MS20030, NLS.

11. Hamlin, *Public Health*, 74–83. See also Comrie, *Scottish Medicine*, 610–11; B. White, "Scottish Doctors and the English Public Health," in *The Influence of Scottish Medicine*, ed. Derek A. Dow, 79–88 (Park Ridge, N.J.: Parthenon Publishing Group, 1988), 82–83; and Pelling, *Cholera, Fever*, 41, 45. On divisions among Evangelicals, see Hilton, *Age of Atonement*, 62–63, 87, 108.

12. *Memoir*, 42.

13. Perkin, *Rise of Professional Society*, 29.

14. Lambert, *John Simon*, 245.

15. Dorothy E. Watkins, *The English Revolution in Social Medicine, 1889–1911* (Ph.D. diss., University of London, 1984), 32.

16. The discussion and all figures are drawn from Watkins, *English Revolution in Social Medicine*, 58–71.

17. Ibid., 71.

18. On 13 July 1863 JBS noted "3–5.15 with Gh[etal] at Vestry preparing Tables." MS ADD 179/19, UCL.

19. See, for example, 17–23 February 1861, MS ADD 179/17, UCL.

20. Anne Hardy outlined its aims in "Public Health and the Expert: The London Medical Officers of Health, 1856–1900," in *Government and Expertise: Specialists, Administrators and Professionals, 1860–1919*, ed. Roy Macleod, 128–42 (Cambridge: Cambridge University Press, 1988), 130–33; and Anne Hardy, *The Epidemic Streets: Infectious Disease and the Rise of Preventive Medicine, 1856–1900* (Oxford: Clarendon Press, 1993), esp. 4–7.

21. My account here is drawn from the extant reports: "Sanitary Report for the Year 1856" NLM and Marylebone Library, London; "Sanitary Report for the Year 1857," GLRO; "Report on the Health of Paddington, during Quarter Ending Michaelmas, 1861," GLRO; "Report . . . , during Half-Year Ending Lady-Day, 1862," Marylebone Library; "Report . . . during Quarter Ending Midsummer, 1862," Marylebone; "Report . . . during Quarter Ending Michaelmas, 1862," Marylebone; "Report . . . during the Half-Year Ending Lady-Day, 1863," Marylebone; "Report . . . during the Half-Year Ending Lady-Day, 1864," Marylebone; "Report during Half-Year Ending Michaelmas, 1865," Marylebone; "Sanitary Report for the Year, 1866–67," NLM, all by JBS.

22. JBS, "Sanitary Report for the Year 1856," 8, 7. St. Mary's population increased from seventeen thousand in 1851 to twenty-six thousand in 1861. St. John's population was constant at twenty-nine thousand. (I am rounding to the nearest thousand.)

23. In 1865 the population of St. Mary's was thirty-nine thousand and of St. John's thirty-seven thousand.

24. This discussion of the sanitary life in Paddington draws on Richard Evans, *Death in Hamburg: Society and Politics in the Cholear Years, 1830–1910* (London: Penguin Books, 1987), chap. 2: "The Urban Environment"; JBS's reports; *Memoir;* and Wohl, *Endangered Lives*, chap. 4: "The Valley of the Shadow of Death."

25. [13 November 1866], PC8/119, PRO, "Extract from Mr. Simon's First Annual Report to the City Commissioners of Sewers," 6 November 1849.

26. Wohl, *Endangered Lives*, 81, and quoted in Evans, *Death in Hamburg*, 127.

27. JBS, "Report during Half-Year Ending Michaelmas, 1865," 18.

28. *Memoir*, 43.

29. JBS, "Sanitary Report for the Year 1856," 11–12.

30. Ibid., 14–15, 46.

31. Ibid., 20.

32. Watkins, *English Revolution in Social Medicine*, 45.

33. Ibid., 46.

34. Hardy, "Public Health and the Expert," 138, makes a similar point.

35. JBS, "Report . . . during Quarter Ending Michaelmas, 1862," 1.

36. Hardy, *Epidemic Streets*, discusses these issues in detail, see esp. 267–69.

37. JBS, "Sanitary Report for the Year 1856," 18.

38. Christopher Hamlin, "Providence and Putrefaction," *Victorian Studies* 28 (1985): 382.

39. Ibid., 384–86; and Hamlin, *Science of Impurity*, 130–31.

40. Hamlin, "Providence and Putrefaction," 390.

41. John Simon to JBS, copy of letter of March 3 [1857], fol. 30, MS20032, NLS.

42. This is a theme throughout Lambert, *John Simon*. See also Simon, "Experiment as a Basis of Preventive Medicine," *Public Health Reports* (London: Offices of the Sanitary Institute, 1887).

43. Simon, *Observations Regarding Medical Education; in a Letter, Addressed to the President of the Royal College of Surgeons* (London: Henry Renshaw, 1842), 6, 8.

44. Simon, *General Pathology, as Conducive to the Establishment of Rational Principles for the Diagnosis and Treatment of Disease; A Course of Lectures, Delivered at St. Thomas's Hospital, during the Summer Session of 1850* (Philadelphia: Blanchard and Lea, 1852), 12.

45. See Simon to JBS, 2 February [1859, copy], fols. 20–21, MS20032, NLS.

46. Simon to JBS, 5 June [1859], fol. 31, MS20032, NLS.

47. See PC8/19, PRO, letter of Simon, 10 December 1861.

48. JBS to GBS, 24 September 1859, fol. 13, MS ADD 179/97, UCL.

49. JBS to GBS, [13 September 1860?], fols. 18–20, MS ADD 179/97, UCL.

50. *Memoir*, 46.

51. H. Weber to GBS, 1 November 1906, MS ADD 179/92; and MS ADD 179/77, UCL.

52. *Memoir*, 49–51, 56.

53. [@1858–60], fol. 23, MS ADD 179/97, BSC, UCL, bracketed notes are GBS's.

54. *Memoir*, 51.

55. Ibid.

56. JBS to GBS, [1859?], fols. 15–16, MS ADD 179/97, UCL.

57. See Hardy, *Epidemic Streets*, 295, for a discussion of the problems associated with nineteenth-century mortality statistics.

58. JBS, "Report . . . during Quarter Ending Michaelmas, 1861," 1.

59. JBS, "Report . . . during quarter ending Michaelmas, 1862," 1.

60. Hardy, *Epidemic Streets;* the quotation is on 294.

61. JBS, "Sanitary Report for the Year 1856," 18–20. Hamlin, *Science of Impurity,* discusses a subject I am here avoiding, the actual purity or safety of the water.

62. For example, Evans, *Death in Hamburg,* uses British cities to illustrate the relative lack of attention to public health by the public officials of Hamburg. Hardy, *Epidemic Streets,* traces an improvement in health beginning in the 1870s due to the diminishing effects of infectious diseases.

63. Hamlin, *Public Health,* 2. Hamlin also problematizes the definition of public health, 1–15.

64. JBS, "On Cerebro-Spinal Meningitis about the Lower Vistula," *Eighth Report of the Medical Officer of the Privy Council* (1865).

Chapter Three: Before the Germ Theory

1. Peter Alter, *The Reluctant Patron: Science and the State in Britain, 1850–1920,* trans. Angela Davis (New York: Berg, 1987), 1, 13.

2. Ibid., 67–69. His source was *Nature* 94 (1914–15): 552–53.

3. Simon's surveys resembled the 1837–38 Fever Investigations, mentioned in his book *Sanitary Institutions* (Pelling, *Cholera, Fever,* 35–36;. Lambert, *John Simon,* 318). See, for example, JBS's report "On Diphtheria at Waltham Abbey," *Eighth Report of the Medical Officer of the Privy Council* (1865), 251–52.

4. On Simon's relationship with the Treasury, see PC 8/33, PC8/51, and PC8/116, PRO. For the quotation, see Simon to Treasury, 14 November 1864, PC 8/51, PRO.

5. GBS to Jane Burdon Sanderson, 12 February 1863, fol. 123, MS ADD 179/97, UCL.

6. "Instructions to Dr. Whitley to go to S. Petersburg and Report" (three sheets unsigned, dictated by Simon), 6 April 1865, PC8/52, PRO.

7. For the aims of the inquiry, see Simon's memorandum of 19 April 1865, PC8/52, PRO. JBS, "On Cerebro-Spinal Meningitis about the Lower Vistula," 287, 268.

8. The first quotation is from JBS to Jane Burdon Sanderson (Jenny), 7 November 1865, fol. 41, MS ADD 179/97, UCL. The second from JBS to Jane Burdon Sanderson, April or May 1866, fol. 141, MS ADD 179/97, UCL.

9. Most of the information about the cattle plague epidemic is drawn from the: *First Report with Minutes of Evidence, and Appendix,* 1865; *Second Report with Minutes of Evidence, and Appendix,* 1866; *Third Report with Appendix, 1866, British Parliamentary Papers,* Reports from Commissioners, vol. 22, sess. 1866.

10. William Bulloch, *The History of Bacteriology* (London: Oxford University Press, 1938), provides an overview of the emergence of the field of bacteriology. See also James Trostle, "Early Work in Anthropology and Epidemiology: From Social Medicine to the Germ Theory, 1840 to 1920" in *Anthropology and Epidemiology,* ed. Craig R. Jones et al. (Boston: Kluwer Academic Publishers, 1986), 35–57. For an account of the British context, see Keith Vernon, "Pus, Sewage, Beer, and Milk: Microbiology

in Britain, 1870–1940," *History of Science* 28 (1990): 289–325. On the American context, see Nancy Tomes, *The Gospel of Germs: Men, Women, and the Microbe in American Life* (Cambridge: Harvard University Press, 1998).

11. John R. Fisher, "British Physicians and the Cattle Plague, 1865–66," *Bulletin of the History of Medicine* 67 (1993): 651–69, notes that a minority of commissioners represented the landed interests (661).

For information about the commissioners, see, for (1) Henry Bence Jones (1814–73), *Dictionary of National Biography (DNB)*, ed. Leslie Stephen and Sidney Lee (1891–92; rpt., Oxford: Oxford University Press [OUP], 1963–65), 10:998–99; (2) Richard Quain (1816–98, not to be confused with his elder cousin Richard Quain [1800–1887], the anatomist), O'Conner, *Founders of British Physiology*, 88; (3) Thomas Wormald (1802–73), *DNB*, OUP rpt. of 1900 ed., 21:945–46; (4) Edmund Alexander Parkes (1819–76), *A Manual of Practical Hygiene*, 6th ed. (New York: W. Wood, 1883); (5) Charles Spooner (1806–71), *DNB*, OUP rpt. of 1897–98 ed., 18:815–16; (6) Robert Ceely, Lambert, *John Simon*, 316; (7) Lyon Playfair (1818–98), French, *Antivivisection and Medical Science*, 73; (8) John Poyntz Spencer, fifth Earl Spencer (1835–1910), *DNB* 2d supp., ed. Sidney Lee (London: Smith, Elder, and Co., 1912), 3:369–72; (9) Robert Arthur Talbot Gascoyne Cecil, Viscount Cranborne (1830–1903), *DNB* 2d supp., 1:329–43; (10) Robert Lowe, later Viscount Sherbrooke (1811–92), *DNB*, OUP rpt. of 1893 ed., 12:197–201; (11) Clare Sewell Read (1826–1905), *DNB* 2d supp., 3:168–69. I have no further information about John Robinson M'Clean.

12. Fisher, "Cattle Plague," 660–61; and Lambert, *John Simon*, 296.

13. On veterinary medicine and its relationship to medicine, see Lise Wilkinson, *Animals and Disease: An Introduction to the History of Comparative Medicine* (Cambridge: Cambridge University Press, 1992). For contemporary medical coverage of the cattle plague epidemic, see in *Lancet*, 6 January 1866, 21–22, the letters of Charles Murchison, JBS, and E. A. Parkes under the title "The Points of Resemblance between Rinderpest and Small-Pox"; and C.B. Redcliffe's letter on "The Treatment of the Cattle Plague," 22–23. For more letters, see *Lancet*, 20 January 1866, 79–80. The *British Medical Journal* summarized "The Appendix to the Third Report of the Cattle-Plague Commission," 14 July 1866, 42–43. In its regular section "Specimens from the Lower Animals" the *Transactions of the Pathological Society of London* 17 (1866): 441–65, presented a series of reports on the cattle plague, which included Dr. Crisp's report (448).

14. *First Report*, 41–42. See also Michael Worboys, "Germ Theories of Disease and British Veterinary Medicine, 1860–1890," *Medical History* 35 (1991): 308–27, esp. 309–10 and 318–19, where he discusses how livestock diseases in Britain were viewed as foreign.

15. *First Report*, vii, xxiii, and xi.

16. Pelling, *Cholera, Fever*, 16–19, 298–99, 302; E. H. Ackerknecht, "Anticontagionism between 1821 and 1887," *Bulletin of the History of Medicine* 22 (1948): 562–93; and Roger Cooter, "Anticontagionism and History's Medical Record," *The Problem of Medical Knowledge: Examining the Social Construction of Medicine*, ed. Peter Wright and Andrew Treacher (Edinburgh: Edinburgh University Press, 1982), 87–108.

17. *First Report*, xiv.

18. Fisher, "Cattle Plague," 661 n. 42.

19. JBS noted reading the debate on the Reform Bill, 13 April 1867, MS ADD 179/23, UCL.

20. *First Report*, 40–42.

21. Ibid., 44.

22. Ibid.

23. Pelling, *Cholera, Fever,* 120–21; and Hamlin, *Science of Impurity,* 130, discuss Liebig's theory and its influence in Britain. See also Justus von Liebig, *Agricultural Chemistry, or Organic Chemistry in Its Application to Agriculture and Physiology,* edited from the manuscript of the author by Lyon Playfair (Philadelphia: T. B. Peterson, 1841); and the discussion in chap. 2.

24. *Oxford English Dictionary,* 2d ed., 832; *Third Report,* vii.

25. Hamlin, "Providence and Putrefaction," 384–86.

26. Eyler, *Victorian Social Medicine,* 101–3.

27. *First Report,* 42; see also Lambert, *John Simon,* 400–401.

28. *Third Report,* iii. See also Worboy, "Germ Theories of Disease," 314–15.

29. *Third Report,* xiv.

30. Lionel S. Beale, *How to Work with the Microscope,* 4th ed. (1857; rpt., London: Harrrison, Paul, 1868); and Beale, *Bioplasm: An Introduction to the Study of Physiology and Medicine* (London: J & A Churchill, 1872).

31. JBS to Jane Burdon Sanderson, 7 November 1865, MS ADD 179/97, fols. 41–42, UCL.

32. R.W. Allen, *Practical Vaccine Treatment for the General Practitioner* (London: H. K. Lewis and Co., 1919), 1–2.

33. *Third Report,* iv.

34. Ibid.

35. Ibid.

36. Lionel S. Beale, "Microscopical Researches on the Cattle Plague," *Third Report,* 129.

37. *Third Report,* v–vi.

38. Beale, "Microscopical Researches," 133.

39. Ibid., 151.

40. Ibid.,150

41. "The Appendix to the Third Report of the Cattle-Plague Commission," *British Medical Journal,* 14 July 1866, 42.

42. Ibid., 42–43, on JBS's research; and 43 for the comments on Beale's conclusions.

43. Gerald L. Geison, "Louis Pasteur," *Dictionary of Scientific Biography,* 10:383. See also JBS, "Bacteria," *Nature,* 29 November 1877, 84.

44. On discipline formation, see John W. Servos, *Physical Chemistry from Ostwald to Pauling: The Making of a Science in America* (Princeton: Princeton University Press, 1990); Geison and Holmes, *Research Schools;* Robert E. Kohler, *From Medical Chemistry*

to Biochemistry: The Making of a Biomedical Discipline (Cambridge: Cambridge University Press, 1982, esp. intro. chap.; and the classic Gerald Lemaine et al., eds., *Perspectives on the Emergence of Scientific Disciplines* (The Hague: Mouton, 1976).

45. *Oxford English Dictionary*, compact ed. (Oxford: Oxford University Press, 1971), 2:243.

46. Inquirer, "The Germ Theory and Spontaneous Generation," *Contemporary Review* (April 1877): 911.

47. *Third Report*, vi.

48. Fisher, "Cattle Plague," 663, 665.

49. Pelling, *Cholera, Fever*, 116. Pelling also points out that between 1830 and 1860 theories of epidemic disease were based on a pathology of the body fluids, and in particular of blood (14–16); on the continuity in sanitary practices, see 294. For a critical appraisal of the reception of Listerian theory and practice, see Christopher Lawrence and Richard Dixey, "Practicing on Principle: Joseph Lister and the Germ Theories of Disease," in *Medical Theory, Surgical Practice*, ed. Christopher Lawrence, 153–215 (New York: Routledge, 1992).

50. Lambert, *John Simon*, 399–401.

51. Ibid., 401.

52. Ibid., 379–80.

53. See the letters from Simon of 18 December 1873 and 16 March 1874, which included his budget for research, in PC8/201, PRO.

54. MS ADD 179/75, UCL.

55. Simon, 13 December 1865, PC8/79, PRO, on the grant to the Royal Geographical Society; see Alter, *Reluctant Patron*, 88.

56. Alter, *Reluctant Patron*, 71.

57. The first quotation is from the letter of 27 March 1874 from William Law of the Treasury Chambers to the clerk of the Privy Council, PC8/201, PRO. See also Simon to the president, Privy Council, 27 March 1876, PC8/212, PRO. On the savings, see John Lambert to Treasury, 23 March 1876, PC8/213, PRO. For Simon's remarks, see his letter to Treasury, 31 March 1874, PC8/201, PRO. See the long correspondence over the 1876–77 grant, PC8/212, PRO; and George Buchanan, *Annual Report of the Medical Officer of the Local Government Board*, 1878, prepared on behalf of Dr. Seaton, the new medical officer of the local government board.

58. Lambert, *John Simon*, 401–2.

59. E. A. Schäfer to JBS, 19 December 1875, fols. 107–10, MS20501, NLS.

60. JBS to Simon, 17 December 1874, fols. 88–89, MS20501, NLS; see also draft letter, JBS to Simon, n.d., fols.198–99, MS20502, NLS.

Chapter Four: From Clinician-Researcher to Professional Physiologist

Epigraph: JBS, diary, 1861, MS ADD 179/17, UCL; and diary, 1866, MS ADD 179/25, UCL.

1. George Harley, Charles Handfield Jones, John Whitaker Hulke, William M. G.

Hewitt, and John Burdon Sanderson, eds., *The New Sydenham Society's Year-Book of Medicine, Surgery, and the Allied Sciences for 1860* (London: New Sydenham Society, 1861).

2. JBS, "Croonian Lecture on the Influence Exerted by the Movements of Respiration on the Circulation of the Blood," *British Medical Journal*, 13 April 1867, 411–13.

3. *Memoir;* and an autobiographical note, MS ADD 179/74, UCL. The quotation comes from JBS, notes of GBS from letters and diaries, bk. 1, 30 April 1866, fol. 48, MS ADD 179/97, UCL.

4. GBS to Jane Burdon Sanderson, 8 May 1866, fol. 145, MS ADD 179/97, UCL.

5. Jacalyn Duffin, *To See with a Better Eye: A Life of R. T. H. Laennec* (Princeton: Princeton University Press, 1998), 286–88.

6. JBS, address to the Middlesex Hospital Medical School, 1 October 1868, MS ADD 179/114, UCL.

7. JBS, "Introductory Address on the Motives and Methods of Medical Study, Delivered at the Commencement of the Winter Session at University College," *Lancet*, 9 October 1880, 566.

8. JBS, "On the Study of Physiology," 212.

9. JBS, diary, 1864, MS ADD 179/20, UCL. See "The Sphygmograph in English Medical Practice," *Lancet*, 26 May 1866. This article and JBS's use of the sphygmograph are also discussed extensively by Robert Frank in "The Telltale Heart: Physiological Instruments, Graphic Methods, and Clinical Hopes, 1854–1914," 211–90, in *Investigative Enterprise*, 211–12, 219. See also P. R. Fleming, *A Short History of Cardiology* (Amsterdam: Wellcome Institute Series in the History of Medicine, Rodopi: 1997), 115–19.

10. Frank, "Telltale Heart," 215–18; see also Audrey B. Davis, *Medicine and Its Technology: An Introduction to the History of Medical Instrumentation* (Westport: Greenwood Press, 1981), 126–28; Stanley Joel Reiser, *Medicine and the Reign of Technology* (Cambridge: Cambridge University Press, 1978), 101–4; and C. Keith Wilbur, *Antique Medical Instruments* (West Chester, Pa.: Schiffer Publishing, 1987), 29.

11. Frank, "Telltale Heart," 215–18. For short biographies of Ludwig and Marey, see Rothschuh, *History of Physiology*, 204–9, 274.

12. JBS, *Handbook of the Sphygmograph* (London: Hardwicke, 1867). The lecture had also been printed, under a slightly different title, in the *British Medical Journal*, 13 July 1867, 12–22; 20 July 1867, 39–40; 27 July 1867, 57–58.

13. JBS, *Handbook of the Sphygmograph*, chap. 1.

14. Ibid., 13–19.

15. On the history of cardiology generally, see Fleming, *Short History of Cardiology;* James B. Herrick, *A Short History of Cardiology* (Springfield: Charles C. Thomas, 1942); and Joshua O. Leibowitz, *The History of Coronary Heat Disease* (Berkeley: University of California Press, 1970). Quain was widely cited: Fleming, 160–76; Herrick, 216; Leibowitz, 113–14; "On Fatty Diseases of the Heart," appeared in *Medico-Chirurgical Transactions* 33 (1850): 121–96. As I noted in chapter 3, it is sometimes difficult to distinguish the two Richard Quains, who were cousins.

16. JBS, "Mode and Duration," *Handbook of the Sphygmograph*, 53.

17. Ibid., 52. On the heart sounds, see Jacalyn Duffin, "The Cardiology of R. T. H.

Laennec," *Medical History* 33 (1989): 42–71. On Williams, see Fleming, *Short History of Cardiology*, 77, 90–94; Reid was not mentioned.

18. Fleming, *Short History of Cardiology*, 81. Duffin, *To See with a Better Eye*, discusses the British style, see esp. 188; and Duffin, "The Cardiology of R. T. H. Laennec." On Corrigan, see Herrick, *Short History of Cardiology*, 124–26; Ralph H. Major, *Classic Descriptions of Disease; with Biographical Sketches of the Authors*, 3d ed. (1932; rpt., Springfield: Charles C Thomas, 1978), 352–57; Frederick A. Willius and Thomas E. Keys, *Classics of Cardiology* (originally titled *Cardiac Classics*) (1941; rpt., New York: Dover Publications, 1961), 419–40, which includes a reprint of Corrigan, "On Permanent Patency of the Mouth of the Aorta, or Inadequacy of the Aortic Valves," *Edinburgh Medical and Surgical Journal* 37 (1832): 225–45.

19. Jean Nicolas Corvisart, *An Essay on the Organic Diseases and Lesions of the Heart and Great Vessels* (1806), trans. by Jacob Gates (1812; rpt., New York: Hafner Publishing Co., 1962), 286.

20. JBS, *Handbook of the Sphygmograph*, 19.

21. JBS, "Mode and Duration," *Handbook of the Sphygmograph*, 55–56.

22. JBS, *Handbook of the Sphygmograph*, 45–47.

23. Corvisart, quoted by Duffin, *To See with a Better Eye*, 178, see also 179; and Leibowitz, *History of Coronary Heart Disease*, 104–5. Again, JBS did not refer to Burns.

24. JBS, "Mode and Duration," *Handbook of the Sphygmograph* 57–64; and JBS, *Handbook of the Sphygmograph*, 21–31.

25. Ibid., 31–33.

26. JBS, "Mode and Duration," *Handbook of the Sphygmograph*, 67–69. On heart failure, see Fleming, *Short History of Cardiology*, 144, 31,145.

27. JBS, "Mode and Duration," *Handbook of the Sphygmograph*, 66.

28. Ibid., 65–66.

29. Duffin, *To See with a Better Eye*, 179–80; and Leibowitz, *History of Coronary Heart Disease*, 104.

30. This account is based on JBS's diaries; and JBS, "The Influence Exerted by the Movements of Respiration on the Circulation of the Blood," *British Medical Journal*, 13 April 1867, 411–13.

31. JBS, "The Influence Exerted by the Movements," 411; and JBS, *Handbook of the Sphygmograph*, 36, in which the Croonian lecture is cited. Barbara Bates, *A Guide to Physical Examination and History Taking*, 4th ed. (Philadelphia: Lippincott, 1987), 265.

32. JBS, "Influence Exerted by the Movements," 413.

33. Ibid, 412.

34. Fleming, *Short History of Cardiology*, 109, citing Christopher Lawrence, "Physiological Apparatus in the Wellcome Musuem," 2. "The Dudgeon Sphymograph and Its Descendants," *Medical History* 23 (1979): 109.

35. Fleming, *Short History of Cardiology*, 182.

36. Ibid., 144.

37. Ibid., 106.

38. Frank, "Telltale Heart," 222–25.

39. On private research support in this era, my source was Alter, *Reluctant Patron,* 82; for the members of the committee, see *Memoir,* 82.

Chapter Five: Becoming a Research Pathologist

Epigraph: GBS to Jane Burdon Sanderson, 10 April 1865, fols. 124–25; attributed to JBS, n.d., 1866, fol. 46, "Notes from Letters and Diaries by Lady Burdon Sanderson," MS ADD 179/97, UCL.

1. There is a large literature on the history of bacteriology; see, for example: Bulloch, *History of Bacteriology;* Trostle, "Early Work in Anthropology and Epidemiology"; Vernon, "Pus, Sewage, Beer, and Milk"; Tomes, *Gospel of Germs.* See also James Strick, foreword to Thomas D. Brock, *Robert Koch: A Life in Medicine and Bacteriology,* vii–xix, 1st ed. (1988; rpt., Washington, D.C.: American Society for Microbiology Press, 2000); and the special edition on "Rethinking the Reception of the Germ Theory of Disease: Comparative Perspectives," *Journal of the History of Medicine and Allied Sciences* 52 (1997).

2. The references in notes 1 and 13 of the introduction provide an overview of the history of physiology. On the nature of scientific controversies and how scientists garner support from their peers, see David Gooding and Trevor Pinch, eds., *The Uses of Experiment: Studies in the Natural Sciences* (Cambridge: Cambridge University Press, 1989).

3. Lawrence and Dixey, "Practicing on Principle," outline the contemporary reconstruction of the history of Listerian surgery. They also emphasize the multiplicity of germ theories, see esp. 156.

4. Simon, *Tenth Report of the Medical Officer of the Privy Council* (1867), 19; and JBS, "On the Communicability of Tubercle by Inoculation," *Tenth Report,* 136.

5. JBS, "On the Communicability of Tubercle by Inoculation," 111, 116–47.

6. Ibid., 150; and JBS, "Further Report on the Inocubality and Development of Tubercle," *Eleventh Report of the Medical Officer of the Privy Council* (1868), 117.

7. Simon, *Twelfth Report of the Medical Officer of the Privy Council* (1869), 58, 60.

8. JBS, "Introductory Report on the Intimate Pathology of Contagion," *Twelfth Report,* 229.

9. Victor A. McKusick, "Jean-Baptiste Auguste Chauveau," in *Dictionary of Scientific Biography,* 3:219–20.

10. JBS, "Introductory Report on the Intimate Pathology of Contagion," 254, 244–45.

11. Ibid., 247. On Hallier, see Jean Theodorides, "Ernst Hans Hallier," in *Dictionary of Scientific Biography,* 6:72–73.

12. JBS, "Introductory Report on the Intimate Pathology of Contagion," 255.

13. JBS, "Further Report of Researches Concerning the Intimate Pathology of Contagion," *Thirteenth Report of the Medical Officer of the Privy Council* (1870), 48, 58, 86. Theodorides, "Hallier," 73, discusses the objections to Hallier's assertions.

14. JBS, "Further Report of Researches Concerning the Intimate Pathology of Contagion," *Thirteenth Report,* 62, 65, 67.

15. JBS, "On the Process of Fever," *Reports of the Medical Officer of the Privy Council and*

Local Government Board, n.s., no. 6 (1875): 9, 46, 30. Russell C. Maulitz, "Rudolf Virchow, Julius Cohnheim and the Program of Pathology," *Bulletin of the History of Medicine* 52 (1978): 162–82, discusses pathology research in this era.

16. JBS, "Reports of an Experimental Study of Infective Inflammations," *Reports of the Medical Officer of the Privy Council and Local Government Board,* n.s., no. 6 (1875): 47–68 (report of 1872 on 69–79; report of 1875 on 47–49). My discussion of the surgical distinction is based on Lawrence and Dixey, "Practicing on Principle," 8–15, MS.

17. JBS, "Reports of an Experimental Study of Infective Inflammations, 49–50; my emphasis.

18. Ibid., 51.

19. Simon, *Report of the Medical Officer of the Privy Council and the Local Government Board,* n.s., no. 6 (1875): 4.

20. Ibid., 4–5.

21. JBS, "Reports of an Experimental Study of Infective Inflammations," 69.

22. Ibid., 79.

23. JBS, "Report of a Further Investigation of the Properties of the Septic Ferment," *Report of the Medical Officer of the Privy Council and Local Government Board,* n.s., no. 8 (1876): 11–12.

24. Simon, *Report of the Medical Officer,* n.s., no. 8 (1876): 1. On Panum, see Rothschuh, *History of Physiology,* 271, 334–35; Charles-Edward A. Winslow, *The Conquest of Epidemic Disease* (Princeton: Princeton University Press, 1943), 267–70; and Bulloch, *History of Bacteriology,* 388. On Panum's research with putrid liquids, see JBS, "Report of a Further Investigation of the Properties of the Septic Ferment," 11–12.

25. JBS, "Report of a Further Investigation of the Properties of the Septic Ferment," 12–13.

26. JBS "Criticisms of Dr. Chauveau of Lyons on the Discussion at the Pathological Society on Pyaemia," *British Medical Journal,* 26 October 1872, 459–60.

27. *British Medical Journal,* "Reports of Societies: Clinical Society of London, Friday, March 13th, 1874"; the discussion was on "Pyæmia in Private Practice," 21 March 1874, 380–86; JBS spoke on the subject, 383–84.

28. JBS, quoted in *Lancet,* "Medical Societies: Pathological Society of London," 10 April 1875, 512.

29. See the letters from Pettenkofer, 24 February 1870, fols. 12–13; 2 March 1870, fol. 16; 26 September 1870, fols. 17–18, the illegible letter from Billroth, 30 March 1870, fols. 14–15; and letter from Chauveau, 17 July 1871, fols. 31–32, all in MS ADD 179/1, UCL.

30. H. Charlton Bastian, "An Address on the Germ Theory of Disease; Being a Discussion of the Relation of Bacteria and Allied Organisms to Virulent Inflammations and Specific Contagious Fevers," *Lancet,* 10 April 1875, 501.

31. Ibid. See also "Discussion on the Germ Theory of Disease, April 6th 1875," *Transactions of the Pathological Society of London* 26 (1875): 255–345; and the account of the debate in *Lancet,* "Medical Societies: Pathological Society of London," 10 April 1875, 511. John Drysdale, M.D., *The Germ Theories of Infectious Disease* (London: Bail-

liere, Tindall and Cox, 1878); Drysdale was identified on the title page as president of the Literary and Philosophical Society of Liverpool and author of "The Protoplasmic Theory of Life."

32. JBS, quoted in "Medical Societies: Pathological Society of London," 512.

33. These remarks summarizing JBS's position in 1875 are drawn from JBS, "Lectures on the Occurrence of Organic Forms in Connection with Contagious and Infective Diseases," *British Medical Journal*, 16 January 1875, 69–71; 13 February, 199–201; 27 March, 403–5; 3 April, 435–37.

34. Ibid., 16 January 1875, 71.

35. See the drafts of letters from JBS to Tyndall, MS ADD 179/3: 20 May 1877, fols. 1–2, and 21 May 1877, fols. 3–4; and Tyndall's reply of 21 May 1877, fols. 5–6, UCL. See also Tyndall, "Note on Dr. Burdon Sanderson's Latest Views of Ferments and Germs," *Proceedings of the Royal Society of London*, 21 June 1877, 353–57; and JBS, "Remarks on the Attributes of the Germinal Particles of *Bacteria*, in Reply to Prof. Tyndall," *Proceedings of the Royal Society of London*, 22 November 1877, 416–26.

36. Gerald L. Geison, "Lionel Smith Beale," *Dictionary of Scientific Biography*, 1:540; and the Beale obituaries: *British Medical Journal*, 7 April 1906, 836–37; and *Lancet*, 7 April 1906, 1004–7.

37. Inquirer, "The Germ Theory and Spontaneous Generation," 902; for the description of JBS, see 916.

38. Pelling, *Cholera, Fever*, 77–78; and J. K. Crellin, "The Problem of Heat Resistance of Micro-Organisms in the British Spontaneous Generation Controversies of 1860–1880," *Medical History* 10 (1966): 50–59, makes this point as well (50). On spontaneous generation, see James Strick, "Darwinism and the Origin of Life: The Role of H. C. Bastian in the British Spontaneous Generation Debates, 1868–1873," *Journal of the History of Biology* 32 (1999): 1–42; James Strick, "Purity and Contamination: John Tyndall, H. Charlton Bastian, Louis Pasteur and Spontaneous Generation, 1870–1878," in *Pasteur, Germs, and the Bacteriological Laboratory*, ed. B. Fantini and J. Buchwald (Cambridge, Mass. MIT Press, forthcoming); James Edgar Strick, "The British Spontaneous Generation Debates of 1860–1880: Medicine, Evolution, and Laboratory Science in the Victorian Context" (Ph.D. diss., Princeton University, 1997); John Farley, *The Spontaneous Generation Controversy from Descartes to Oparin* (Baltimore: Johns Hopkins University Press, 1977), chaps. 5 and 7; Geison, *The Private Science of Louis Pasteur*, chap. 5, esp. 128–29; John Farley and Gerald L. Geison, "Science, Politics and Spontaneous Generation in Nineteenth-Century France: The Pasteur–Pouchet Debate," *Bulletin of the History of Medicine* 48 (1974): 161–98. On spontaneous generation and Darwin, which Strick discusses, see also W. F. Bynum, "Darwin and the Doctors: Evolution, Diathesis, and Germs in Nineteenth-Century Britain," *Gesnerus* 40 (1983): 43–53; and J. K. Crellin, "The Dawn of the Germ Theory: Particles, Infection and Biology," in *Medicine and Science in the 1860s*, ed. F. N. L. Poynter (London: Wellcome Institute of the History of Medicine, 1968), 57–76.

39. Strick, "Darwinism and the Origin of Life"; and see his discussion in "British Spontaneous Generation Debates," 216–45.

40. Simon to JBS, 22 July 1877, MS ADD 179/3, fols. 15–16, UCL.

41. JBS, "Lectures on the Infective Processes of Disease," *British Medical Journal,* 22 December 1877, 880. The lectures continued 29 December 1877 (913–15); 5 January 1878 (1–2); 12 January (45–47); 26 January (119–20); 9 February (179–83).

42. Dougall quoted in "Discussion on the Germ Theory," 297.

43. Drysdale, *Germ Theories,* 8–9.

44. JBS, "Introductory Report on the Intimate Pathology of Contagion," *Twelfth Report of the Medical Officer of the Privy Council* (1869): 232.

45. Beale to JBS, 16 November 1870, MS ADD 179/1, fols. 24–25, UCL.

46. Tyndall's letters to T.H. Huxley, vol. 8: 18 November 1876, fol. 197; 14 December 1877, fols. 198–99; 14 January 1877, fol. 200; 8 December 1880, fol. 227, all in T.H. Huxley Papers, ICL.

47. Strick discusses in detail the medical community's reaction to Tyndall, "The British Spontaneous Generation Debate," 216–45; he provides outlines of its reaction in "Purity and Contamination"; and "Darwinism and the Origin of Life."

48. Beale to JBS, 14? November 1870, MS ADD 179/1, fols. 22–23, UCL.

49. Huxley to Tyndall, 18 November 1876, T. H. Huxley Papers, ICL.

50. JBS, "Further Report of Researches Concerning the Intimate Pathology of Contagion," *Thirteenth Report,* 48–69.

51. Alter, *Reluctant Patron,* 100.

52. Dougall quoted in "Discussion on the Germ Theory of Disease," 296; my emphasis.

53. Beale, "Dr. Sanderson's Speech on the Germ Theory of Disease," *Lancet,* 17 April 1875, 558–59.

54. James Lambert, *The Germ Theory of Disease, a Paper Read before the Irish Central Veterinary Medical Society, April 5, 1883* (London: Bailliere, Tindall and Cox, 1883), 4.

55. T. Maclagan, M.D., *The Germ Theory Applied to the Explanation of the Phenomena of Disease: The Specific Fevers* (London: Macmillan, 1876), 32, 21. The booklet was dedicated to John Tyndall.

56. Bastian, "Address on the Germ Theory of Disease," 502.

57. Bastian, *The Beginnings of Life* (New York: Appleton, 1872), 2:cxx, app. E, quoted in Crellin, "Dawn of the Germ Theory," 72. Crellin also noted "the attractive simplicity" of the germ theory.

58. JBS, "Lectures on the Infective Processes," esp. 181–82. For an account of Koch's researches, see Brock, *Robert Koch,* chap. 5.

59. See also Lawrence and Dixey, "Practicing on Principle"; and A.J. Youngson's discussions of the debate about Listerism and "the germ theory," *The Scientific Revolution in Victorian Medicine* (New York: Holmes and Meier, 1979), 194–99.

60. JBS to Simon, draft letters (28 April and 2 May 1876) in reply to "Letters of Instruction," fols. 116–77, MS20501, NLS.

61. First quotation is from 1 December; second from 9 December 1878, "Notes Taken from Burdon Sanderson's Diaries, Scrapbooks, Etc.," MS ADD 179/107, UCL.

Chapter Six: Focusing on Physiology

Epigraph: Dracula was first published in 1897.

1. Letter 1:8, 15 August 1873, Darwin-JBS Correspondence, UBC.

2. Letter 1:14, 9 September [1873], Darwin Burdon Sanderson Correspondence, UBC. Printed in Frances Darwin, ed., *More Letters of Charles Darwin* (London: John Murray, 1903), 2:394.

3. JBS, "On the Electrical Phenomena Which Accompany the Contractions of the Leaf of Dionæa muscipula," *British Association Reports* 43 (1873): 133.

4. JBS, MS ADD 179/74,UCL.

5. For current representations of "Action Potentials in Higher Plants," including the *Mimosa* and Venus's-flytrap, see Bertil Hille, *Ionic Channels of Excitable Membranes* (Sunderland, Mass.: Sinauer Associates, 1984), 377.

6. See Coleman's discussion of the cell theory, *Biology in the Nineteenth Century*, 23.

7. Grmek, *Le Legs de Claude Bernard*, chap. 9: "La Fleur au bois dormant." This research was not well known (332).

8. JBS, envelope, "Papers Connected with Address in Physiology," July 1873, *Omitted Pages*, fols. 17–18, MS ADD 179/72, UCL.

9. George Buckmaster to JBS, 20 January 1895, fol. 37; and Schäfer to JBS, 27 January 1896, fols. 105–6, MS ADD 179/7, UCL.

10. Hacking, *Representing and Intervening*, 133–34, defined representation.

11. Fritz Kurtz, "Zur Anatomie des Blattes der Dionæa muscipula," *Archiv für Anatomie, Physiologie und Wissenschaftliche Medizin* (Leipzig) (1876): 1–29.

12. Frank, "Telltale Heart," 227; he provides an abbreviated account of these experiments (23–35). On the graphic method in physiology, see the work of Merriley Borell, esp. "Extending the Senses: The Graphic Method," *Medical Heritage* 2 (1986): 39–50; and "The Kymograph and the Origins of the Graphic Method," *Electrical Quarterly* 7 (1985): 2–3.

13. JBS, "Note on the Electrical Phenomena Which Accompany Irritation of the Leaf of *Dionæa muscipula*," *Proceedings of the Royal Society, London* 21 (1873): 495–96. Reports of the lecture are: JBS, "On the Mechanism of the Leaf of *Dionæa Muscipula*, and on the Electrical Phenomena Which Accompany Its Contraction," *Royal Institution Proceedings* 7 (1875): 332–35. JBS, "Venus's Fly-trap (*Dionæa muscipula*)," *Nature* 10, 11 June 1874, 105–7; 18 June 1874, 127–28, is a more complete account of the same lecture.

14. Ludimar Hermann, *Elements of Human Physiology*, ed. and trans. Arthur Gamgee (London: Smith, Elder and Co., 1878), 290.

15. Frank, "Telltale Heart," 230.

16. Ibid., 226–27.

17. Ibid., 230.

18. Ibid., 232–33.

19. JBS and F. J. M. Page, "Experimental Results Relating to the Rhythmical and Excitatory Motions of the Ventricle of the Heart of the Frog, and of the Electrical Phenomena Which Accompany Them," *Proceedings of the Royal Society London* 27, 23 May 1878, 410–14; JBS and F. J. M. Page, "On the Time Relations of the Excitatory Process in the Ventricle of the Heart of the Frog," *Journal of Physiology* 2 (1880): 384–435; and JBS and F. J. M. Page, "Notice of Further Experimental Researches on the Time-Relations of the Excitatory Process in the Ventricle of the Heart of the Frog," *Proceedings of the Royal Society London* 30, 13 May 1880, 373–83.

20. JBS and Page, "On the Time Relations of the Excitatory Process in the Ventricle of the Heart of the Frog"; and JBS and Page, "Notice of Further Experimental Researches on the Time-Relations of the Excitatory Process in the Ventricle of the Heart of the Frog." The two papers were similar and utilized the same rheotome of JBS's design. The *Journal of Physiology* article was a more detailed account.

21. Fig. 6.4 is from JBS, "On the Time Relations of the Excitatory Process in the Ventricle of the Heart of the Frog," 393. "Figure 1," in JBS and Page, "Notice of Further Experimental Researches on the Time-Relations of the Excitatory Process in the Ventricle of the Heart of the Frog" (375), is almost the same. Fig. 6.4 pictured here has the addition of the K1 circuit.

22. JBS and Page, "On the Time Relations of the Excitatory Process in the Ventricle of the Heart of the Frog," 393.

23. JBS, "On a New Rheotome," *Proceedings of the Royal Society* (13 May 1880): 383–87; he and Page also described the rheotome in an appendix to "On the Time Relations of the Excitatory Process in the Ventricle of the Heart of the Frog," see app. 2, 432–35. On Hermann and Bernstein, see Rothschuh, *History of Physiology,* 222, 217.

24. JBS, "On the Time Relations of the Excitatory Process in the Ventricle of the Heart of the Frog," 426–29.

25. JBS, "The Excitability of Plants," *Proceedings of the Royal Institution* 10 (1882): 147. See also Hermann Munk, "Die elektrischen und Bewegungs-Ersheinungen am Blatte der *Dionæa muscipula,*" *Archiv für Anatomie, Physiologie und Wissenschaftliche Medizin (Reichert's and Du Boise-Reymond's Archiv)* (1876): 30–122; and A. J. Kunkel, "Electrische Untersuchungen an pflanzlichen und thierischen Gebilden," *Archiv für die Gesammte Physiologie* 25 (1881): 342–79.

26. JBS, "On the Electromotive Properties of the Leaf of *Dionæa* in the Excited and Unexcited States," *Philosophical Transactions of the Royal Society London* 73 (1881): 7.

27. This was evident in the German papers that JBS arranged to have translated into English: JBS, ed., *Translations of Foreign Biological Memoirs,* vol. 1: *Memoirs on the Physiology of Nerve, of Muscle and of the Electrical Organ* (Oxford: Clarendon Press, 1887). In particular, see Emil DuBois-Reymond, "On Secondary Electromotive Phenomena of Muscle," trans. Mrs. Lauder Brunton, 163–228; and Hermann, "On So-Called Secondary Electromotive Phenomena in Muscle and Nerve," trans. Francis Gotch, 277–330.

28. For one of many letters about purchasing instruments abroad, see JBS to Schäfer, 6 May [1876], PP/ESS/B7/2, WL.

29. Ludwig to JBS, 20 July 1877, fols.13–14, MS ADD 179/3, UCL. Ludwig wrote about an upcoming visit to London, when he would stay with JBS.

30. GBS to [Mary Elizabeth Haldane], 19 June 1870, MS ADD 179/115, UCL.

31. Schäfer to JBS, 2 May 1872, fols. 31–32, MS20501, NLS. JBS to Schäfer, [1875], PP/ESS/B7/5; and 6 May [1876], PP/ESS/B7/2, WL.

32. See JBS's letters from Charles Beevor while Beevor was in Germany: 20 May 1883, fols. 80–81; and 2 July 1883, fols. 82–83, MS20030, NLS.

33. Envelope "Papers Connected with Address in Physiology," *Omitted Pages*, July 1873, MS ADD 179/72, UCL.

34. Emil DuBois-Reymond to JBS, 6 June 1877, fols. 8–10, MS ADD 179/3, UCL.

35. Huxley to JBS, 14 June 1877, fols. 11–12, MS ADD 179/3, UCL.

36. JBS, MS ADD 179/115, UCL.

37. GBS, 8 July 1879, MS ADD 179/115, UCL. See also GBS's note of a return visit to Professor Stricker, in Vienna; Stricker had previously stayed with the Sandersons in London, 10 October 1877, MS ADD 179/115, UCL.

38. GBS, 21 April 1880, MS ADD 179/115, UCL.

39. JBS to Schäfer, 1 May 1880, PP/ESS/B8/1, WL.

40. [Fleische], 26 July 1879, fols. 41–42, MS ADD 179/3, UCL. Fleische and Lippmann had been responsible for the modifications of the electrometer.

41. Christian Scøven to JBS, 10 March 1881, fol. 9, MS ADD 179/4, UCL.

42. See Gustav Fritsch's letter asking JBS for introductions to use in Ceylon (1 September 1904, fols.137–38, MS20031, NLS), which JBS provided; see the letter from someone in Ceylon who agreed to facilitate Fritsch's visit (fols. 204–5, MS 20031, NLS). JBS and DuBois-Reymond corresponded on many occasions; see, for example, DuBois-Reymond to JBS: 8 August 1884, fols. 204–5, MS ADD 179/4; 28 January 1885, fols. 109–10, MS ADD 179/5; 18 July 1888, fols. 138–40, MS ADD 179/5, UCL.

43. JBS, "Opening Address to Biology Section of Instrument Exhibition," *Nature*, 1 June 1876, 117–19.

44. Thomas Buzzard to JBS, 15 May 1879, fols. 35–36, MS ADD 179/3, UCL.

45. JBS, "Opening Address to Biology Section of Instrument Exhibition." The preceding quotations in this section are from this address.

46. On British positivism, see Martha S. Vogeler, *Frederic Harrison: The Vocations of a Positivist* (Oxford: Clarendon Press, 1984); and Terence R. Wright, *The Religion of Humanity: The Impact of Comtean Positivism on Victorian Britain* (Cambridge: Cambridge University, 1986). See also the discussion of Victorian positivism in Sandra den Otter, *British Idealism and Social Explanation: A Study in Late Victorian Thought* (Oxford: Oxford University Press, 1996), 52–53.

47. Hacking, *Representing and Intervening*, 42.

48. J. S. Haldane to L. K. Trotter, 3 December 1891, Mitchison Papers, quoted by Sturdy, Steven Waite Sturdy, "A Co-Ordinated Whole: The Life and Work of John Scott Haldane" (Ph.D. diss., University of Edinburgh, 1987), 183. Sturdy discussed how these philosophical differences affected J. S. Haldane and JBS's views of each other's research (182–83).

49. Frederic Harrison cited by den Otter, *British Idealism*, 2, quoting Vogeler, *Frederic Harrison*, 195. On Victorian idealism, see den Otter, *British Idealism;* for the Haldanes' particular brand, see Sturdy, "A Co-ordinate Whole," esp. 17–27, 36–46.

50. Sturdy, "Co-Ordinated Whole," 20.

51. Georg Wilhelm Friedrich Hegel, *Lectures on the Philosophy of Religion, Together with a Work on the Proofs of the Existence of God*, trans. E. B. Speirs and Jane Burdon Sanderson, 3 vols. (London: Kegan Paul, Trench, Trübner, and Co., Ltd., Paternoster House, Charing Cross Road, 1895). Speirs completed it after her death.

52. T. N. Stokes, "A Coleridgean against the Medical Corporations: John Simon and the Parliamentary Campaign for the Reform of the Medical Profession, 1854–8," *Medical History* 33, no. 3 (1989): 343–59.

53. JBS, "Our Duty to the Consumptive Bread-Earner," address to the Oxford and District and Reading and Upper Thames Branches of the British Medical Association, 1901, reprinted in the *Memoir*, 300–301.

54. For a definition of research schools, see Servos, "Research Schools and Their Histories," 10; and Geison, "Scientific Change, Emerging Specialties, and Research Schools," chart 2, 24, both of whom drew on J. B. Morrell's classic article "The Chemist Breeders"; see also the discussion in the next chapter.

55. Edward Sharpey-Schafer, *History of the Physiological Society during Its First Fifty Years, 1876–1926* (supp. to *Journal of Physiology* [December 1927]) (London: Cambridge University Press, 1927), 24. In the wake of anti-German sentiments during World War I and in honor of his predecessor William Sharpey, Schäfer changed his surname to Sharpey-Schafer.

56. E. Ray Lankester to JBS, fols. 198–99, n.d., MS20031, NLS.

57. Ritchie to JBS, 21 May 1901, fol. 70, MS ADD 179/9, UCL; for Dunstan, who was also a close friend of JBS, see MS ADD 179/14, UCL.

58. See O'Connor, *Founders of British Physiology*, 55–57; and Sharpey-Schafer, *History of the Physiological Society*, 31. See also the correspondence between JBS and Stricker about the matter, January and February 1871, fols. 16–20, MS20501, NLS.

59. GBS, MS ADD 179/107, UCL.

60. See, for example, Schäfer to JBS, 19 December 1875, fols. 107–10; and 16 February 1873, fols. 41–42, MS20501, NLS. Here Schäfer asked JBS for advice about ongoing research and publications. There was extensive correspondence between the two men.

61. For information about Gotch's career, see O'Connor, *Founders of British Physiology*, 149–50; and Sharpey-Schafer, *History of the Physiological Society*, 66, 151. See also Gotch to JBS, 3 April 1883, fols. 137–40; 6 April 1883, fols. 141–44; 13 April 1883, MS ADD 179, UCL.

62. Gotch to JBS, 8 May 1892, fols. 79–82, MS ADD 179/6, UCL.

63. There are six other men who in their correspondence acknowledge JBS's assistance with either experiments or their careers whom I have not further identified. They are Alfred M. Gossage, 28 August 1888, fols. 144–46, MS ADD 179/5, UCL; M. Armand Thuffer? 19 May 1891, fols. 131–32, MS20501, NSL; Richard Harding

Notes to Pages 131-135 ... 209

Bremridge, 3 February 1894, fols. 155–57, MS20501, NLS; A. A. Kanthack, 21 July 1892, fols. 100, and 2 August 1899, fol. 101, MS ADD 179/6, UCL; Bertram Hunt, n.d. [1895], fols. 49–50, MS ADD 179/7, UCL; G. G. Stakes, 5 June 1873, fols. 63–64, MS20501.

64. Charles E. Beevor to JBS, 3 June 1899, fols. 7–8, MS179/8, UCL.

65. Horsley to JBS, 10 November, 1896, fols. 132–33, MS ADD 179/7, UCL.

66. Titchener to JBS, 21 June 1895, fol. 84, MS ADD 179/7, UCL.

67. Mott to JBS, 7 June 1899, fols. 50–51, MS ADD 179/8, UCL.

68. See, for example, JBS to Schäfer, PP/ESS/B7/17, WL.

69. John Abel to Mary Hinman, 25 February 1883, Abel Papers, record group 1, related in John Parascandola, *The Development of American Pharmacology* (Baltimore: Johns Hopkins Press, 1992), 25 and 158 n. 12.

70. [?] to GBS, 19 June 1906, MS20033, NLS.

71. Quoted in Strick, "British Spontaneous Generation Debates," 271.

72. Geison, *Michael Foster*, 358.

73. Schäfer to JBS, 4 August 1894, fols. 154–55, MS ADD 179/6, UCL. See also JBS to Schäfer, 31 July [1894], fols. 151–52, MS ADD 179/9, UCL; and JBS to Schäfer, 6 August [1894], MS ADD 179/13, UCL.

74. JBS, "On the Electromotive Properties of the Leaf of *Dionæa* in the Excited and Unexcited States—Second Paper," *Philosophical Transactions of the Royal Society* 179B (1888): 417–49; and see JBS and F. J. M. Page, "On the Electrical Phenomena of the Excitatory Process in the Heart of the Frog and of the Tortoise, as Investigated Photographically," *Journal of Physiology* 4 (1883–84): 327–38, for details of the photographic apparatus.

75. Gerald L. Geison, "The Protoplasmic Theory of Life and the Vitalist-Mechanist Debate," *Isis* 60 (1969): 272.

76. Ibid., 273–74, 279.

77. My analysis of the articles cited in the classic work by Francis Ernest Lloyd, *The Carnivorous Plants* (1942; rpt., New York: Dover Publications, 1976), demonstrated a substantial increase in the number of publications on this group of plants in the 1870s and 1880s. See also Grmek, *Le legs de Claude Bernard*, 317; and Huxley, "On the Borderline Territory between the Animal and the Vegetable Kingdoms," lecture at the Royal Institution, 28 January 1876 (*McMillan's Magazine*, 1874), reprinted in Huxley, *Science and Culture and Other Essays* (London: Macmillan, 1882).

78. Robert Bentley Todd and William Bowman, *The Physiological Anatomy and Physiology of Man* (Philadelphia: Blanchard and Lea, 1857), wrote an extensive introduction (fifty-one pp.), which included such topics as "Life" and "Theories of Life" and discussed animals and plants together.

79. Michael Foster, *A Text Book of Physiology* (London: Macmillan, 1877), 1.

80. E. A. Schäfer, *Text-book of Physiology*, vol. 1 (Edinburgh: Young J. Pentland, 1898).

81. Julius von Sachs, *Lectures on the Physiology of Plants*, trans. H. Marshall Ward (Oxford: Clarendon Press, 1887), 650. Sachs supported the conclusions of Kunkel,

who had worked in his laboratory. Note also that the botanist Sydney Howard Vines, who became JBS's colleague at Oxford, made no mention of JBS's research in his botany textbook, *Lectures on the Physiology of Plants* (Cambridge: Cambridge University Press, 1886).

82. See the discussion in Bruno Latour, "Drawing Things Together," in *Representation in Scientific Practice,* ed. Michael Lynch and Steve Woolgar, 19–68 (Cambridge, Mass.: MIT Press, 1990).

83. Sachs, *Lectures,* 650.

84. DuBois-Reymond, "Observations and Experiments on Malapterurus," trans. Edith Prance, in *Translations of Foreign Biological,* 1:369.

85. JBS, "Croonian Lecture on the Relation of Motion in Animals and Plants to the Electrical Phenomena Which Are Associated with It," *Proceedings of the Royal Society London,* 16 March 1899, 63; and W. Biedermann, *Electro-Physiology,* vol. 2, trans., Frances A. Welby (London: Macmillan, 1898), chap. 6: "Electromotive Action in Vegetable Cells."

Chapter Seven: Physicians, Antivivisectionists, and Oxford Physiology

Epigraph: JBS, draft letter, [1882], fols. 71–72, MS ADD 179/4, UCL.

1. See, for example, Acland to JBS, 18 June 1882, fols. 55–56, MS ADD 179/4, UCL; and James B. Atlay, *Sir Henry Wentworth Acland: A Memoir* (London: Smith Elder and Co., 1903). On the Waynflete chair, see Margaret Pelling, "The Refoundation of the Linacre Lectureships in the Nineteenth Century," in *Linacre Studies,* ed. F. R. Maddison, M. Pelling, and C. Webster (Oxford: Clarendon Press, 1977), 266–67.

2. JBS, "Written 'n the Train' 1882," fol. 25, MS20032, NLS, handwritten copy, duplicated in part, MS ADD 179/114, UCL.

3. Huxley to Brodrick (warden of Merton College), in *Life and Letters of T. H. Huxley,* ed. Leonard Huxley (New York: D. Appleton and Co., 1900), 2:32.

4. "The Oxford Professorship of Physiology," *British Medical Journal* 2 (1882): 956, quoted in Stella V. Butler, "Centers and Peripheries: The Development of British Physiology, 1870–1914," *Journal of the History of Biology* 21 (1988): 485.

5. For one of many long and exceedingly plaintive letters in which Gamgee (1841–1905) outlines his unhappiness in Manchester and his hopes of filling the Oxford position, see Gamgee to JBS, 5 February 1881, fols. 3–8, MS ADD 179/4, UCL.

6. Lankester to JBS, 8 July [1882], fols. 59–60; see also W. T. Thiselton Dyer (1843–1928) to JBS, 10 July 1882, fols. 61–62. For a later example, see Ernest Hart (1835–98), editor of the *British Medical Journal,* 7 November [1882], fols . 75–77, all in MS ADD 179/4, UCL.

7. M'Kendrick to JBS, 13 July 1882, fols. 63–64, MS ADD 179/4, UCL.

8. Acland to JBS, 17 July 1882, fols. 65–68, MS ADD 179/4, UCL.

9. Price to JBS, 19 July 1881, fols. 51–52, MS20030, NLS.

10. See the correspondence in MS20030, NLS, fols 53–57 and 62–65; for the notice of JBS's election, see fols. 86–87, MS ADD 179/4, UCL.

11. Acland to JBS, 25 November 1882, fols. 90–95, MS ADD 179/4, UCL.

12. Huxley to JBS, 27 November 1882, fols. 98–99, MS ADD 179/4, UCL.

13. On the British medical profession, see the references in intro. n. 4; and Irvine Loudon, *Medical Care and the General Practitioner, 1750–1850* (Oxford: Clarendon Press, 1987); and Christopher Lawrence, *Medicine in the Making of Modern Britain* (London: Routledge, 1994).

14. See the discussion and references in chap. 6 n. 55; see also George Weisz, *The Emergence of Modern Universities in France, 1863–1914* (Princeton: Princeton University Press, 1983); Tuchman, *Science, Medicine, and the State in Germany;* and Kohler, *From Medical Chemistry to Biochemistry.*

15. Stella V. F. Butler, "Centers and Peripheries," table 2, 478.

16. Geison, "Divided We Stand," 68; Shortt, "Physicians, Science and Status"; and, for a different view, see John Parascandola, "The Search for the Active Oxytocic Principle of Ergot: Laboratory Science and Clinical Medicine in Conflict," 205–27, in *Neue Beiträge zur Arzneimittelgeschichte: Festchrift für Wolfgang Schneider zum 70. Geburtstag,* ed. Erika Hickel and Gerald Schröder (Stuttgart: Wissenschaftliche Verlagsgesellschaft MBH, 1982).

17. See, for example, Warner, "Science in Medicine," esp. 45; John Harley Warner, "Ideals of Science and Their Discontents in Late Nineteenth-Century American Medicine," *Isis* 82 (1991): 454–78; Steve Sturdy, "The Political Economy of Scientific Medicine: Science, Education and the Transformation of Medical Practice in Sheffield, 1890–1922" *Medical History* 36 (1992): 125–59; John Harley Warner, "Science, Healing, and the Physician's Identity: A Problem of Professional Character in Nineteenth-Century America," in *Essays in the History of Therapeutics,* ed. W. F. Bynum and V. Nutton, special issue of *Clio Medica* 22 (1991): 65–88, esp. 67–68; and Lawrence, "Incommunicable Knowledge," 503–4.

18. Lawrence makes a similar point in "Incommunicable Knowledge." See also Lawrence, *Medicine in the Making of Modern Britain,* 68–72.

19. Bonner, *Becoming a Physician,* 205.

20. For a discussion of medical character and gender, see Gert H. Brieger, "Classics and Character: Medicine and Gentility," *Bulletin of the History of Medicine* 65 (1991): 94; and Rebecca Tannenbaum, "Earnestness, Temperance, Industry: The Definition and Uses of Professional Character among Nineteenth-Century American Physicians," *Journal of the History of Medicine and Allied Sciences* 49 (1994): 262–70, 276–82. See also Peterson, *Medical Profession in Mid-Victorian London,* 179–86, in which she discusses how issues of class and expertise contributed to a dispute between nurses and doctors at Guy's Hospital.

21. See the references in intro. n. 14.

22. JBS to Jane Burdon Sanderson (his sister), Good Friday, 1882, bk. 2, fol. 170, in *Extracts of Letters, Diaries* made by GBS, Burdon Sanderson Collection, UBC.

23. "Rusticus" in a letter to the editor also noted that the current system penalized science and mathematics students *(Oxford Magazine,* 4 February 1885, 49–50).

24. Janet Howarth, "Science Education in Late-Victorian Oxford: A Curious Case

of Failure?" *English Historical Review* 102 (1987): 335; and n. 3. The figures for Cambridge cited by Howarth were taken from R. M. MacLeod and R. C. Moseley, "The Anatomy of an Elite: The Natural Science Trips and Its Graduates, 1850–1914" (SSRC report, 1976).

25. Howarth, "Science Education," 359: "Cambridge, after 1860, permitted candidates who had passed the Previous to proceed direct to the NST."

26. G. C. Bourne, letter to the editor, "Science and Pass Mods," *Oxford Magazine*, 18 February 1885, 89.

27. Huxley, 21 December 1885, *Life and Letters*, 2:127.

28. Howarth, "Science Education," 342 states that half of those who read the Natural Sciences Tripos between 1886 and 1914 became doctors, quoting R. N MacLeod and R. C. Moseley, "The 'Naturals' and Victorian Cambridge," *Oxford Review of Education* (1980): 186. Mark Weatherall, "Scientific Medicine and the Medical Sciences in Cambridge, 1851–1939" (Ph.D. diss., University of Cambridge, 1994), charts the percentages of physiology students at Cambridge who were medical students and concludes "that the proportion is unlikely to have been less than 50%, and was very probably more" (67–68).

29. See table 1, "Oxford and Cambridge Medical Degrees Awarded, 1801–1900," in Alastair H. T. Robb-Smith, "Medical Education in Nineteenth-Century Oxford," in *Nineteenth Century Oxford*, pt. 1 (*History of University of Oxford*, vol. 6), ed. M. G. Brock and M. C. Curthoys (Oxford: Oxford University Press, 1998).

30. [William] B. Carpenter to Acland, 5 May 1885, MS. Acland d. 92, fols. 49–54, Bodl. I gratefully acknowledge the permission of the Bodleian Library to quote from the Acland papers.

31. Richard Symonds, *Oxford and Empire: The Last Lost Cause?* (London: Macmillan, 1986), "Note on Oxford Terminology," xvii–xviii. For a detailed discussion of the fellows in this period, see A. J. Engel, *From Clergyman to Don: the Rise of the Academic Profession in Nineteenth Century Oxford* (New York: Oxford University Press, 1983).

32. Engel, *From Clergyman to Don*, 57.

33. Ibid., 60–61.

34. Ibid., 27–33.

35. Letter of [25 July 1869], in *Dear Miss Nightingale: A Selection of Benjamin Jowett's Letters to Florence Nightingale, 1860–1893*, ed. Vincent Quinn and John Prest (Oxford: Clarendon Press, 1987), 173.

36. On Congregation in this period, see Engel, *From Clergyman to Don*, 199–200. JBS to E. A. Schäfer, 6 August [1894], MS ADD 179/13, UCL.

37. See French, *Antivivisection and Medical Science*, 275–76 n. 170, for a brief account of the events at Oxford. See also E. B. Nicholson's letters to JBS: 19 October 1883, fols. 167–68; and 20 October 1883, fol. 169, MS ADD 179/4, UCL. And see subsequent discussion of JBS's worries about Freeman's opposition in the 1890s.

38. Mary Ann Elston, "Women and Anti-vivisection in Victorian England, 1870–1900," in Rupke, *Vivisection in Historical Perspective*, 259–94, see esp. 282; and Coral

Lansbury, *The Old Brown Dog: Women, Workers, and Vivisection in Edwardian England* (Madison: University of Wisconsin Press, 1985).

39. GBS, 27 February, 1881? fol. 204, GBS's notes, bk. 2, Burdon Sanderson Collection, UBS.

40. *Gardeners' Chronicle,* 19 June 1875, 790, from the *Pall Mall Gazette,* 16 June 1875, 5.

41. See French, *Antivivisection and Medical Science,* chap. 9, "The Mind of Antivivisection: Medicine," esp. 330, 332.

42. Quoted in French, *Antivivisection and Medical Science,* 276, see also 275. JBS, *Extracts* by GBS, bk. 3, 29, Burdon Sanderson Collection, UBC.

43. For a discussion of these issues, see "Notes and News," *Oxford Magazine,* 29 April 1885, 179–83; and "Moderations," *Oxford Magazine,* 28 October 1885, 322–23. See also the letter of William Benjamin Carpenter to Acland (cited in n. 30), which stressed the literary abilities of his scientifically trained son and his attainment of "general culture."

44. JBS, 6 November 1885, bk. 3, 50, *Extracts,* by GBS, Burdon Sanderson Collection, UBC.

45. Engel discussed Freeman's views in *From Clergyman to Don.* It was a narrow defeat; the preamble fell by one vote, 71 to 72. Although the traditionalists won, it was a small victory, since the bill was defeated over confusion about the exact wording of the preamble (*Oxford Magazine,* 6 May 1885, 209–10). The announcement of the final changes appeared in the *Oxford Magazine,* 11 November 1885, 357.

46. *Oxford Magazine,* 2 December 1885, 411.

47. "Notes and News," *Oxford Magazine,* 16 June 1886, 252.

48. Paul Willert to Poulton, 25 June 1885, fol. 43, MS ADD 179/5, UCL.

49. Howarth, "Science Education," 355, 357; and see her table 4.

50. Engel, *From Clergyman to Don.*

51. I thank Mark Curthoys for this point (pers. comm.).

52. Donald MacAlister to JBS, 3 November 1884, fol. 217, MS ADD 179/4, UCL. JBS to Henry Pitman, secretary, Royal College of Physicians, 8 November 1884, fols. 218–19, MS ADD 179/4, UCL. Pitman to JBS: 11 November 1884, fols. 220–21; and 12 November 1884, fols. 222–23, MS ADD 179/4, UCL. Although I assume JBS also consulted the Royal College of Surgeons, I have found no record of a reply.

53. Robb-Smith, "Medical Education in Nineteenth-Century Oxford."

54. MacAlister to JBS, 13 November 1884, fols. 227–28, MS ADD 179/4, UCL.

55. Pitman to JBS, 14 November 1884, fols. 229–30, MS ADD 179/4, UCL.

56. I have here condensed the many possible paths to an Oxford or Cambridge medical degree. For fuller particulars of the requirements as they stood in 1885, see *British Medical Journal,* 13 September 1884, 507–8; Robb-Smith, "Medical Education in Nineteenth-Century Oxford"; and Weatherall, "Scientific Medicine and the Medical Sciences." On medical education generally, see Charles Newman, *The Evolution of Medical Education in the Nineteenth Century* (London: Oxford University Press, 1957); and Bonner, *Becoming a Physician.*

57. Frances Henry Champneys to Acland, 30 January 1885, fols. 55–58, MS Acland d.92, Bodl. Champneys had matriculated in 1866 at Brasenose College; he received a B.A. degree in 1870 and M.A. and M.B. degrees in 1875.

58. *Lancet* noted the letter to the *Pall Mall Gazette* in order to chastise the writer and indicate its support of the reform proposal (21 February 1885).

59. This controversy emerged from letters to JBS and Acland and memoranda generated by the Oxford Medical Graduates Club. See MS. Acland d. 92, Bodl.: Champneys to Acland, 30 January 1885, fols. 55–58; G. Humphry to Acland, 2 February 1885, fols. 66–67. From MS ADD 179/5, UCL: J. Matthews Duncan to JBS, 30 January 1885, fols. 3–4; Samuel West to JBS, 27 February 1885, fols. 6–9; Champneys to JBS, 17 February 1885; Perry Kidd to JBS, 4 March 1885, fols. 12–13; West to JBS, 4 March 1885, fols. 14–17; Champneys to JBS, 4 March 1885, fols. 18–21.

60. *Lancet*, 15 March 1884, noted the foundation of the "Oxford Medical Graduates Club" at a meeting at the invitation of Dr. A. B. Shepherd (487). See also *Lancet*, 5 April 1884, 637. *Lancet*, 31 May 1884, noted the inaugural meeting of the club, attended by Acland, JBS, Munro, Church, Corfield, Payne, and about fifty others, including West and John Morgan (995–96).

61. Champneys to Acland, 30 January 1885, fols. 55–58, MS. Acland d.92, Bodl.

62. J. Matthews Duncan to JBS, 30 January 1885, fols. 3–4, MS ADD 179/5, UCL.

63. Humphry to Acland, 2 February 1885, fols. 66–67, MS. Acland d.92, Bodl.

64. Champneys to JBS, 27 February 1885, incomplete letter, fols. 10–11, MS ADD 179/5, UCL.

65. Samuel West to JBS, 27 February 1885, fols. 6–9, MS ADD 179/5, UCL. West had matriculated at Christ Church in 1867; he was granted a B.A. degree in 1871, M.A. and M.B. degrees in 1875, and an M.D. degree in 1882.

66. West to Acland, 6 March [1885], fols. 114–16, MS. Acland d.92, Bodl.

67. "The Oxford Medical School," *British Medical Journal*, 20 February 1886, 357.

68. Champneys to JBS, 4 March 1885, fols. 18–21, MS ADD 179/5, UCL.

69. The motion was carried, 412 votes to 244, *Oxford Magazine*, "The 'Vivisection' Grant," 11 March 1885, 163–64.

70. Canon Liddon, for example, wrote to assure Acland that his speech had not been meant as personal attack (13 March 1885, fols. 24–25, MS. Acland d.98, Bodl.).

71. "Medical Degrees at Oxford," *Lancet*, 13 June 1885, 1092.

72. The relevant papers are in MS ADD 179/66, UCL.

73. Liddell to JBS, 26 April 1885, fol. 35, MS ADD 179/5, UCL.

74. JBS, 29 January 1886, fol. 77, bk. 3, *Extracts*, by GBS, Burdon Sanderson Collection, UBC.

75. JBS to Acland, [1891], fols. 58–59, MS. Acland d.65, Bodl. See also Arthur Thompson (anatomy instructor and eventually professor) to JBS, 25 May 1891, fols. 133–36, MS 20501, NLS; Edward Chapman to JBS, 30 May 1891, fols. 41–42, MS ADD 179/6, UCL; E. Ray Lankester to JBS, 30 May [1891], fols. 139–40, MS 20501, NLS; William Gillem? to JBS, 30 May [1891], fols. 141–42, MS 20501, NLS.

76. See announcements to that effect in *Lancet,* 29 May 1886, 1051; and *British Medical Journal,* 29 May 1886, 1049.

77. A. M. M. Stedman, ed., *Oxford: Its Life and Schools* (London: George Bell and Sons, 1887), ix. Weatherall, "Scientific Medicine and the Medical Sciences," also notes the expenses associated with a Cambridge education (52).

78. Champneys to JBS, 4 March 1885, fols. 18–21, MS ADD 179/5, UCL.

79. Ibid.

Chapter Eight: A Corner Turned?

Epigraph: JBS to F. J. M. Page, 17 January [1895], fol. 14, MS20031, NLS.

1. For Acland's views on medicine, medical education, and the role of science generally, see "Introductory Address Delivered before the Devonshire Association for the Advancement of Science, Literature and Art" (The Association, 1880); and "The Unveiling of the Statue of Sydenham in the Oxford Museum, August 9, 1894" (Oxford: Horace Hart, 1894).

2. Simon to Acland, 1 February 1890, fols. 100–101; and Simon to Acland, 30 March 1890, fols. 102–3, MS. Acland d.92, Bodl.

3. *Oxford Magazine,* 2 November 1892, 46. On its role in the university, see Symonds, *Oxford and Empire,* 9–11.

4. For the table, see "Table 1B: Natural Science Honours Graduates by Subject, 1886–1900," in Howarth, "Science Education," 338; see also 360–62.

5. Lankester to Huxley, 5 December [1888], vol. 21, fols. 123–24, T. H. Huxley Papers, ICL. For the details of Lankester's suit against the vice chancellor, see "Professor Lankester and the Vice-Chancellor," *British Medical Journal,* 24 December 1887; and "The Vice-Chancellor and Professor Ray Lankester," *Oxford Magazine,* 18 January 1888, 153–56. For evidence of Lankester's former disputes, see also the collection of documents at the Zoology Department, Oxford; and in MS ADD 179/68, UCL.

6. Howarth, "Science Education," 362; and JBS, "The School of Medical Science in Oxford" (Oxford, 1892), 28–34.

7. See letters of E. Chapman to JBS, 1 December 1890, fols. 28–29, and 4 December 1890, fls. 32–33, MS ADD 179/6, UCL; and Howarth, "Science Education," 362–63. This move was likely to have Acland's support; see the discussion of his views in Robb-Smith, "Medical Education in Nineteenth-Century Oxford."

8. The dispute emerged clearly in the *Oxford Magazine.* See "Professor's Lankester's Open Letter" of 23 March 1892, which describes his scheme (264–65). See also Lankester's letters to the editor: 30 March 1892, 293–94; 27 April 1892, 312–13, in which he quotes Schäfer to support his position; 1 June 1892, 410, in which he notes JBS's publication of a pamphlet quoting Lankester's own words from the 1870s to support Sanderson's position.

See "Copy of a Circular of the Linacre Professor," 10 April [1892], "Copy of a Letter to Prof. Schäfer," 2 May 1892; and other notes in MS20503, NLS, for JBS's point of view.

Other relevant correspondence includes: JBS to Schäfer, 2 May 1892, PP/ESS/ B8/15, and 11 June 1892, PP/ESS/B8/17; Lankester to Schäfer 12 May 1892, PP/ESS/ B41/10, all in WL. Lankester to JBS, n.d, fols. 192–93, and a partial letter, 16 May [1892], fols. 194–95, MS20031, NLS.

9. Lankester to Acland, n.d., fols.110–12, MS Acland, d.92, Bodl.

10. Acland, "Oxford and Modern Medicine: A Letter to Dr. James Andrew," printed for private circulation (Oxford, 1890). For the correspondence, see the many letters in MS. Acland d.92, Bodl.

11. Church to Acland, 8 February 1892, MS. Acland d. 63, fol. 103, Bodl. quoted in Howarth, "Science Education," 364.

12. JBS to John Cowdry, copy of letter, 13 May 1892, MS20503, NLS.

13. Lankester to JBS, 8 November [@1893], fols. 200–201, MS20031, NLS.

14. Geison, *MichaelFoster*, 335; and see the discussion on 331–55. Geison expressed uncertainty, however, about whether JBS's work was evolutionary; see 361 n. 91.

15. JBS, "The Origin and Meaning of the Term 'Biology,'" presidential address to the British Association, 1893, reprinted in *Memoir*, 242.

16. JBS, "Ludwig and Modern Physiology," lecture delivered at Royal Institution, March 1896, reprinted in *Memoir*, 271.

17. Geison, *Michael Foster*, 332.

18. For example, in the T. H. Huxley Papers (Ill.) there are 6 letters from JBS. This compares with 211 letters from Foster, 30 from George Romanes, 253 from John Tyndall, 60 from Lankester, and 56 from Benjamin Jowett. These numbers are drawn from the catalog of the collection. In all of the JBS papers I have looked at, there are few letters from Huxley and remarkably few from Foster.

19. Rothschuh, *History of Physiology*, 334; for an overview of Chauveau's career, see McKusick, "Jean-Baptiste Auguste Chauveau," in *Dictionary of Scientific Biography*. Panum's career can be pieced together by reading Rothschuh, and Bulloch, *The History of Bacteriology;* and see the discussion in the notes to chap. 5.

20. Duffin, *To See with a Better Eye*, 286–87. Maulitz, "Rudolf Virchow, Julius Cohnheim and the Program of Pathology," in which he notes the physiological influence on Cohnheim, also influenced my analysis.

21. Thomas Henry Huxley, "The Connection of the Biological Sciences with Medicine," address delivered to International Medical Congress in London, 9 August 1881, reprinted in *Science and Culture and Other Essays*, 326.

22. Lawrence, *Medicine in the Making of Modern Britain*, notes, for example, "In fact, by the 1890s, the laboratory's claim to define and establish the presence of febrile disease was relatively uncontested" (73).

23. JBS, "Croonian Lectures on the Progress of Discovery Relating to the Origin and Nature of Infectious Diseases," *Lancet*, 7 November 1891, 1027–32; 14 November, 1083–88; 21 November, 1150–54; 28 November, 1208.

24. Ibid., 1029.

25. Ibid., 1083. On the contrast between German and French styles of bacteriological theory and practice, see Bynum, *Science and the Practice of Medicine in the Nine-*

teenth Century, 160–61; and Paul Weindling, "Scientific Elites and Laboratory Organization in *Fin de Siècle* Paris and Berlin: The Pasteur Institute and Robert Koch's Institute for Infectious Diseases Compared," in Cunningham and Williams, *Laboratory Revolution in Medicine,* 170–88.

26. JBS, "Croonian Lectures on the Progress of Discovery," 1152; my emphasis.

27. Ibid., 1085.

28. Ibid., 1084–88.

29. On the configuration of pathology and bacteriology, see Maulitz, "Pathology," esp. 124.

30. A. B. Griffiths, *Researches on Micro-Organisms Including an Account of Recent Experiments of Microbes in Certain Infectious Diseases—Phthisis, Etc.* (London: Baillière, Tindall and Cox, 1891). See also J. Rosser Matthews, "Major Greenwood versus Almroth Wright: Contrasting Visions of 'Scientific' Medicine in Edwardian Britain," *Bulletin of the History of Medicine* 69 (1995): 30–43, for a discussion of how a "triangular" dispute emerged among clinicians, bacteriologists, and statisticians in Edwardian Britain over who should arbitrate medical knowledge (42).

31. JBS, "Croonian Lectures on the Progress of Discovery," 1151.

32. See "Notes on Discoveries in Bacteriology," n.d. (copy), MS20032, fols.118–20, NLS; and GBS to Ritchie, 15 August 1908 (copy), MS ADD 179/106, UCL.

33. Bremridge to JBS, 3 February 1894, MS20501, fols. 155–57, NLS.

34. *Memoir,* 132–35.

35. Ibid., 63–64.

36. Ibid., 136–37; and JBS, "Our Duty to the Consumptive Bread-Earner," address to the Oxford and District and Reading and Upper Thames Branches of the British Medical Association, 1901, reprinted in *Memoir,* 299–312.

37. This is a recurring theme in Alter, *Reluctant Patron;* see 24, 99, 100, 106–7.

38. JBS, "Croonian Lectures on the Progress of Discovery," 1154.

39. L. E. Shore to JBS, 4 June 1899, fol. 26, MS ADD 179/8, UCL.

40. Foster to JBS, 4 June [1899], fols. 224–25, MS ADD 179/7, UCL.

41. JBS, "The Pathology of Infection," *Transactions of the Pathological Society of London* 54 (1902–3): 5.

42. All the quotations here are from Sherrington's letters to JBS: 17 February 1902, fols. 77–78, 19 February 1902, fols. 81–82, 8 March 1902, fols. 83–86, MS ADD 179/9, UCL. The dean was not named in the letters. According to Thomas Kelly, *For Advancement of Learning: The University of Liverpool, 1881–1981* (Liverpool: Liverpool University Press, 1981), Andrew Melville Paterson was the chair of anatomy from 1894 to 1919 and dean of medicine from 1895 to 1903 (520–21). JBS's protégé Gotch had preceded Paterson as dean from 1892 to 1895.

43. Stella V. F. Butler, *Science and the Education of Doctors in the Nineteenth Century: A Study of British Medical Schools with Particular Reference to the Development and Uses of Physiology* (Ph.D. diss., University of Manchester, 1981), 245.

44. MS20507, NLS

45. *Memoir,* 148; and GBS, 24 November 1905, fol. 109, MS20032, NLS.

Index